OPTIONAL OPTIONS: WORK DESIGN AND MANUFACTURING AUTOMATION

The financial support of the Netherlands Organization for Scientific Research (NWO) for the correction of the English text of *Optional Options: Work Design and Manufacturing Automation*, is gratefully acknowledged.

Optional Options: Work Design and Manufacturing Automation

JOS BENDERS
Department of Businesss Administration
Faculty of Policy Sciences
University of Nijmegen

Avebury
Aldershot · Brookfield USA · Hong Kong · Singapore · Sydney

Published by

Avebury
Ashgate Publishing Limited
Gower House
Croft Road
Aldershot
Hants GU11 3HR
England

Ashgate Publishing Company
Old Post Road
Brookfield
Vermont 05036
USA

British Library Cataloguing in Publication Data

Benders, Jos
Optional Options: Work Design and
Manufacturing Automation
I. Title
670.427

ISBN 1 85628 490 5

Printed and Bound in Great Britain by
Athenaeum Press Ltd., Newcastle upon Tyne.

Contents

List of tables and figures

Preface

This study is the result of many trips, physical as well as intellectual. Its design underwent many changes since its inception in 1988. The contributions of many people proved invaluable for ending successfully.

I am grateful to Job de Haan for his thorough supervision and for forcing me to defend the changes I felt necessary on solid ground. In spite of his incredibly busy affairs, Fred Huijgen never failed to provide critical and judicious review. Without his support, this study would not have been completed. Professor P. Sanders let me profit from his many contacts.

In my view, scientific progress can not be achieved without empirical work. The university as an ivory tower is an undesirable spectre, yet this image is likely without the cooperation of 'the outside world'. Therefore, it is appropriate to thank all persons interviewed for their unselfish cooperation. Within the academic circuit, indulgence in navel-gazing should be avoided. In this respect, Klaas ten Have, Jos Mossink, Niels Noorderhaven, Erik Poutsma and Jannes Slomp helped by providing valuable criticisms on preliminary drafts of various sections. Apart from the persons mentioned above, professors Wout Buitelaar and Arndt Sorge gave useful comments on the complete manuscript. Each in their own way, Robert Evers, Arthur Giesberts, Paul Leermakers and Ivo van der Vegt assisted in various stages of the project.

Furthermore, I am obliged to many friends, at Tilburg University and elsewhere, who let me blew off the steam, which was frequently necessary, especially Henk and Martijn, who did a great job enduring my revelations about the progress. Finally, in a way my parents have laid the foundations for this study by stimulating a continuous process of personal development.

The financial support of the Netherlands Organization for Scientific Research (NWO) for the correction of the English text of *Optional Options: Work Design and Manufacturing Automation*, is gratefully acknowledged.

Permissions acknowledgements

P. 7. Figure from *Sociological Paradigms and Organisational Analysis* by G. Burrell and G. Morgan. Used by permission of G. Burrell and G. Morgan.

P. 8. Quote from 'Sociology of Organizations Around the Globe; Similarities and Differences Between American, British, French, German and Dutch Brands', *Organization Studies*, 11(2) by C.J. Lammers. Used by permission of Walter de Gruyter, Berlin-New York.

Pp. 35-36. Quote from L.E. Davis and J.C. Taylor, 'Technology, Organization and Job Structure', in *Handbook of Work, Organization and Society* (editor R. Dubin). Copyright © by Houghton Mifflin Company.

P. 50. Quote reprinted from *Long Range Planning*, Volume 24, P.T. Bolwijn and T. Kumpe, 'Manufacturing in the 1990s - Productivity, Flexibility and Innovation', Pages 44-57, Copyright 1990, with permission of Pergamom Press Ltd, Headington Hill Hall, Oxford OX3 0BW, UK.

P. 52. Quote from 'The Causal Texture of Organizational Environments', *Human Relations*, 18(1) by F.E. Emery and E.L. Trist. Used by permission of Plenum Publishing Corporation.

P. 53. Quote from 'Environmental Factors and Job and Organization Design: The Case of Olivetti', in *The Quality of Working Life; Volume Two; Cases and Commentary* (editors L.E. Davis and A.B. Cherns) by F. Butera. Used by permission of The Free Press.

P. 64. Tables from 'The training of workers and their occupational level; The qualitative structure of employment in the Netherlands 1960-1985', *Netherlands Journal of Social Sciences*, 26(2) by F. Huijgen. Used with permission of Van Gorcum & Co., Assen.

Pp. 83-84. Quote from *Automation, Production Systems, and Computer Integrated Manufacturing* by M.P. Groover. Used by permission of Prentice-Hall Inc.

P. 96. Data from several *Onderzoeken naar scholings- en arbeidsmarkt-aspecten* published by FME. Used by permission of FME.

Pp. 109-110. Tables from 'Technique, organization, and the future of work in manufacturing' in *New Technologies and the Future of Work* (editor N. van den Heuvel) by E. Poutsma. Used by permission of SISWO.

P. 117. Data from *CNC-machinebedieners, Produktie-organisatie en Flexibele automatisering* by B. Fruytier et al. Used by permission of COB\SER, The Hague.

Pp. 133. Table from *Flexibele automatisering in Nederland; Ervaringen en opinies* by G. Laurentius, H. Timmerman and A.A.M. Vermeulen. Used by permission of Stichting Toekomstbeeld der Techniek, The Hague.

P. 215. Quote from 'Occupational Health and Safety Problems Caused by Increases in the Robot Population', in *Occupational Health and Safety in Automation* (editor K. Noro) by Y. Yokomizo, published in 1986. Used by permission of Taylor and Francis Ltd. London.

1 Prologue

The Roman emperor Vespasian, who ruled from 69 to 79, once rewarded the inventor of a new machine to convey columns, but decided not to use the machine as it would cause unemployment amongst the poor. This example shows that technical inventions, displacement of labor and strategic behavior with respect to these issues date back to the times of the Romans at least. Probably, they are as old as man himself. This can not be said for the systematic study of these phenomena, which started only recently. Although many notions around the theme 'man and machine' can be traced back to long forgone periods, they were not the subject of academic and practitioner investigations until the founders of disciplines such as mechanical engineering and political economy started their work in the eighteenth century. Since then, the study of the increasing division of labor has been subject to an increasing division of labor itself. Disciplines such as ergonomics, psychology, sociology, and industrial engineering gradually became involved in research on the theme, generally without much mutual contact. The growth of the field of researchers went hand in hand with the growth of the field of research. New machines were and are invented continuously. This also holds for organization and management techniques, which are to guarantee that man and machine cooperate productively. Among relatively recent examples of both categories are 'industrial robots' and 'sociotechnical systems theory', which both originated in the 1940s and 1950s. Both have an aura of 'flexibility' and 'smoothness', which have to be seen in relation to developments at product markets and the deployment of labor. The opinions about the factual and desirable paths of development differ on a variety of accounts. This book is a sketch of a map, which hopefully serves to clarify at least part of the landscape formed by man and machine. More concrete: work content around programmable automation forms the topic of study, a topic which can not be studied fruitfully without also examining a score of external factors.

The second chapter sketches the theoretical context of the debate. The first section focuses on work design by juxtaposing two alternative 'design orientations', i.e. the organizing logic underlying the assignments of tasks to functions. These design orientations are called, respectively, 'fractionized' and 'integrative'. Historically, the first approach has been dominant, but during the last forty years a multitude of integrative design methods has been developed, often as a reaction to so-called 'dysfunctions' of fractionized ways of organizing. Are integrative designs actually being implemented to an increasing degree? What are the boons and merits of both design orientations? The next section discusses the impact of technology, or preferable 'technical systems'. These are frequently held responsible for changes in work content, yet the notion that they would determine work content is widely rejected. To what extent are such views justified? Do certain systems more or less automatically lead to specific work designs, or are there 'degrees of freedom', implying that the role of technical systems is limited at most? As organizations do not operate in an environmental vacuum, developments at product and labor markets are discussed in, respectively, the sections 2.3 and 2.4. At product markets, batch sizes are increasing according to some, whereas others point to the need for 'flexibility' has been frequently stressed, requiring an adaptive organization as well. What is the relationship between changes in work content and developments at product markets? Mass education has resulted in a continuously rising level of education at the (supply side of the) labor market. Much literature in the 1980s and early 1990s has suggested that fractionized organizations have become obsolete and need to be converted into integrative ones. Yet, questions abound. What is the descriptive value of this development? How do developments at the labor market influence work design? What alternatives are thinkable? The fifth and final section of chapter two is an attempt to integrate the earlier sections into a research model. The remainder of the study concentrates on the Netherlands in order to hold cultural and institutional factors constant. Hypotheses are formulated, which form the framework for the interpretation and discussion of empirical work in the chapters four and five. First, however, chapter three makes the abstract notion of 'technical system' used in chapter two more concrete by introducing three forms of automation of discrete manufacturing processes: (C)NC-machines, industrial robots and Flexible Manufacturing Systems. Its character is in the first place descriptive. Section 3.1 centres around the description of technical characteristics, among which programming methods, and around different applications of these machines. What is technically feasible? How do technical features relate to economic characteristics? Section 3.2 describes the diffusion of the various forms of automated manufacturing in the Netherlands. Are these of substantial importance, or should they be regarded as exceptional?

Chapter four contains secondary analyses of data, which were gathered in empirical research on work design around the three forms of automated manufacturing discussed in the preceding chapter. The original data are used where possible, and reinterpreted in the light of the hypotheses formulated in section 2.5: how do the factors identified in the theoretical chapter as having an important influence on the division of labor materialize in the case of automated manufacturing? The division of labor around automated manufacturing centres around the assignment of programming tasks to the machine operator, as programming means controlling the machine and the labor process. Much literature is available on (C)NC-machines allowing to discuss the influence of a host of factors. Especially the case data allow to gain a richer insight into the interwovenness of technical and economical factors. Some of these factors simultaneously impede and enhance the occurrence of integrated functions via intermediate factors. Industrial robots and Flexible Manufacturing Systems proved to have been less popular objects of study than (C)NC-machines, although in relation to the actual number of Flexible Manufacturing Systems in use, the number of studies on their impact can be considered to be substantial. Remarkably enough, the industrial robots have received relatively little attention. Therefore, the empirical part of the study focuses on industrial robots, and more specifically on their most popular application: arc welding. Thus, chapter five contains ten case studies, addressing the central question:

> How are the tasks, that can be distinguished around arc welding robots, divided over jobs, and what factors influence the choice for a specific division of these tasks over jobs?

One advantage of these case studies over the secondary analysis of chapter four is that they allow to answer the hypotheses more clearly, thereby also establishing the relative importance of the factors distinguished. But is there a clear pattern discernable? At the end of chapter five, a model is sketched as a more systematic assessment of the role of the actors distinguished.

Finally, chapter six places the previous findings in a historical perspective. The generic theme 'work and technology' has been subject of scientific enquiries for over two centuries now. The outlines of this debate are given, drawing on studies of various periods, from different disciplinary backgrounds, and conducted in different countries. Next, the question is addressed whether or not industrial robots have a favorable impact on the quality of working life, or more specifically, on work content and working conditions. Ultimately, it proves to be possible to relate the empirical findings concerning work content to the historical discussion.

However, before proceeding it seems necessary to pay attention to some epistemological issues in the next section. This has a somewhat contemplative nature, and the reader may feel that it does not belong in the main body of the text. Yet, whereas the epistemological foundations of knowledge are evidently important, their existence is not always realized, let alone the consequences of different epistemological bases. While reviewing theoretical and empirical studies one often touches on irreconcilable findings, which are attributable to diverging epistemological strands. Furthermore, in the remainder of the book occasionally terms are used which are explained in the next section.

1.1 Spectacles and spectacles

Two homonyms of 'spectacles' are 'glasses' and the plural of 'grand performance'. In social science, both homonyms play a critical role. Throughout the centuries, 'technology and labor' have formed the very heart of economic developments, or in more theatrical wordings, they were, and still are, the main actors in the historical play of economic development. Whereas economic historians may be able to write the script of this play after it has been played, the opinions of the audience and the actors themselves may vary, as critics frequently come to different judgements of the same performance. The reason for these differences of opinion is not that the play was different, but that the criteria which the critics use differ. In other words: they wear other spectacles.

A similar process can be seen in research. Authors with various backgrounds have been, are and will be involved in research on the 'technology and labor'. The variety in backgrounds does not only concern the disciplines of authors, but often also their ideological backgrounds and scientific convictions. In 1989 and 1990, conflicting assumptions, reflected in contrasting dominant approaches in literature, formed the (implicitly present) background of a debate between adherents of the labor process approach and proponents of modern sociotechnology in the Dutch/Flemish *Tijdschrift voor Arbeidsvraagstukken*, the 'Journal for Labor Issues'. Earlier, a similar collision had occurred in an article by

Sørensen (1985).

The ideological background of any human being, including academicians, is likely to influence the particular stances taken. The same facts may be interpreted in different ways by different people, depending on the ideological backgrounds of the interpreters. To clarify this statement, the difference between unitary, pluralist and radical approaches will be discussed (Burrell and Morgan, 1979; Morgan, 1986). These approaches will be used to provide a partial explanation of the differing positions taken in the mentioned debate. The aim of this exercise is to enunciate the underlying assumptions of these approaches and to classify various theories in a framework. Before proceeding to this task, a word of warning is necessary: this framework may be interpreted as providing stereotypes of various theories. While such a framework may be used as a useful map to orientate oneself in the world of organizational theory, one must keep in mind that these often are rude descriptions of properties. Members of a population for which a specific category is used, generally possess most, but not necessarily all of the properties mentioned.

The central idea in Burrell and Morgan's 1979 book 'Sociological Paradigms and Organisational Analysis' is, that 'all theories of organization are based upon a philosophy of science and a theory of society. [...] All social scientists approach their subject via explicit or implicit assumptions about the nature of the social world and the way in which it may be investigated' (1979, p. 1). The importance of these statements must not be underestimated. However, many authors seem to be unaware of the implicit assumptions which they have adopted. As stated, the assumptions concern either the scientific approach adopted or a view of society. The scientific assumptions have different natures:

1 ontological: concerning the essence of the phenomena under investigation;
2 epistemological: concerning the grounds of knowledge;
3 concerning human nature, and the relationship of human beings with their environment;
4 methodological: dealing with the way knowledge can be obtained.

Table 1.1 describes two ideal-typical strands on these assumptions as the subjectivist and objectivist approach to social science.

Table 1.1
The subjective-objective dimension

	Subjectivist	**Objectivist**
Ontology	Nominalism	Realism
Epistemology	Anti-positivism	Positivism
Human nature	Voluntarism	Determinism
Methodology	Ideographic	Nomothetic

Source: Adapted from Burrell and Morgan, 1979, p 3

Nominalists assume that there is no real structure to the world external to individual cognition. Social reality is structured by creating concepts and labels which are used to make this world intelligible for human observers. One can be sure of the existence of these concepts, but at the same time the existence of a social world is denied. In contrast, realism postulates the existence of a structured social world, although existing concepts may not suffice to describe this world.

With respect to epistemology, positivists try to explain and predict events in the social world by searching for regularities and causal relationships. By formulating and validating hypotheses and theoretical models, a body of objective knowledge concerning the social world can be construed and extended. Anti-positivists however, maintain that there is no such thing as an 'objective' knowledge. Human behavior and actions can only be understood from the point of view of the actor. It is essential to adopt the actor's frame of reference to be able to understand his actions.

A third dimension of differences between objectivists and subjectivists concerns the issue of human nature. Determinists see man's behavior as being ruled by the environment. Voluntarists on the other hand see man as completely free to pursue his own strategy. Perhaps more than on the other issues, intermediate positions between the extremes of determinism and voluntarism are frequently adopted.

The fourth debate centres around the question how knowledge can be gathered. The positions taken on the three issues mentioned have consequences for the question of what methodology is considered to be applicable. The ideographic approach stresses the importance of getting close to a subject in order to be able to understand the subject's point of view and backgrounds. Impressionistic accounts, like diaries and biographies, provide important data according to the ideographic approach. The nomothetic approach on the other hand stresses systematic protocols and techniques. Key elements include scientific rigour, testing hypotheses and the use of quantitative research methods such as surveys.

With regard to the view of society, Burrell and Morgan make a distinction between 'sociologists of regulation' and 'sociologists of radical change'. Sociologists of the first category tend to emphasise the unity and cohesiveness of society, and are concerned with the need for regulation in human affairs. In contrast, the basic concern of sociologists of radical change is 'to find explanations for the radical change, deep-seated structural conflict, modes of domination and structural contradiction which its theorists see as characterising modern society' (Burrell and Morgan, 1979, p. 17).

By combining the distinctions with regard to society with the distinctions with regard to science, four paradigms of social theories emerge, as shown in Figure 1.1.

SOCIOLOGY OF RADICAL CHANGE

SUBJECTIVIST	Radical humanist	Radical structuralist	**OBJECTIVIST**
	Interpretive	Functionalist	

SOCIOLOGY OF REGULATION

Figure 1.1: Paradigms in social theories

Source: Burrell and Morgan, 1979, p 22

These paradigms provide different viewpoints of the social-scientific reality as they are based on different meta-theoretical assumptions. Most approaches dealt with in this study, and for that matter the study itself, are functionalist, although some radical structuralist theories are used.

While it is beyond the scope of this study to discuss the differences between the various categories extensively, the underlying assumptions of functionalism need discussion in further detail. As shown in Figure 1.1 functionalism is objectivist with respect to its approach to science. Furthermore, functionalism belongs to the sociology of regulation.

Within functionalism, several views may be adopted. To be more concrete, Lammers (1990, pp. 181-3) provides a short history of the concepts 'unitarism' and 'pluralism'. According to him, in 1958 the Englishman Ross was the first one to make a distinction between pluralism and unitarism, using the words 'organic unities' and 'plural societies'. In 1966, Fox suggested to replace the unitary system by a pluralist approach. This advice was followed by many. Lammers himself relabelled the two views as 'socio-cultural system' and 'parties model'.

Burrell and Morgan argue that the interest in power and conflict laid the basis for the analysis of organization as pluralist political systems. Pluralism is the opposite of unitarism. Pluralism and unitarism hold opposing positions with regard to interests, conflict and power. 'Whereas the unitary view of organizations tends to stress that an organization is a co-operative enterprise united in the pursuit of a common goal, the pluralist view stresses the diversity of individual interests and goals' (1979, p. 202). In the unitary view, organization members work together to achieve a goal, from which all organization members benefit. In the pluralist view however, an organization is seen as a network of individuals (or groups of individuals) who only work together to achieve their individual goals. Conflict is thus an inherent property of organizational life: as individual interests may differ, conflict between individuals is always present. However, conflicts may not endanger the organization's survival, and ought to be institutionalized. In contrast, unitarists view conflict as undesirable phenomenon, that should be controlled by managerial actions. Conflicts may disrupt the harmonious order in the organization, and thus prevent the achievement of the organization's goal. The question of power is largely ignored in this view: as all organization members benefit from the organization's well-being, organization members should conform themselves to the rules of their organization as set and advocated by management. 'Concepts such as authority, leadership and control are preferred ways of describing the 'prerogative' of managers to guide the organization as a whole towards desired goals and aims' (1979, p. 203). In the pluralist view however, power is the medium through which conflicts are settled. Organizational life is a never ending power-play between (groups of) individuals.

> Some of these authors also arrive at more than two models because they differentiate between models which can be (and in the sixties and seventies they usually were) considered as variants of either the system view or the parties view' [...] Morgan (1986: Chaps. 6 and 9) subdivides what is called here the parties perspective into the 'pluralist frame of reference' and 'radical organization theory' (Lammers, 1990, p. 183).

Lammers refers to Morgan's book 'Images of Organization', in which this tripartition in organizational theories is made. With this statement, Lammers leaves the underlying view of society (both approaches adopt the same scientific assumptions) out of consideration, as radicalism and pluralism are both seen as the parties perspective. Table 1.2 is a summary of Exhibit 6.6 of Morgan (1986). It shows the main positions taken by the three frames of reference, concerning the topics of 'interests', 'conflict' and 'power'.

Table 1.2
Frames of reference

	Unitarism	**Pluralism**	**Radicalism**
Interests	Common objectives of an integrated team	Recognition of group and individual interests; 'loose coalition'	Interests are largely incompatible; the organization as battleground to achieve own goals
Conflict	Undesired activity of troublemakers, which should be removed	Inherent characteristic or organizational life; potentially positive	Inevitable and part of a class conflict at a societal level
Power	Largely ignored	Crucial medium to resolve or alleviate conflicts	Key feature of organization, but unequally divided following class divisions

Source: Adapted from Morgan, 1986, pp 188-189

An analysis of Table 1.2 leads to the conclusion that pluralism and radicalism can indeed be seen as further subdivisions of the parties view, as Lammers argues. A remarkable difference between pluralism and radicalism is only implicitly being referred to in Table 1.2. This difference concerns the choice for a particular party. While this choice is left open in pluralism, radicalism clearly opts for the underdog, who is in most cases personified as 'worker' in a never ending war with (capitalist) manager. Given the Marxist roots of radicalism, this choice is hardly a surprise. However, one may wonder why these typologies do not contain a category which opts for the perspective of (capitalist) managers. The answer, which is in itself pluralist, is that there is no need for such a separate category, as unitarist theories or their underlying ideologies can be used by managers, or any other leaders for that matter, to serve their own purposes. A unitary frame of reference, leading managers to consciously or unconsciously believe they are managing an organizational team, implies that team members have to gather in the pursuit of common objectives. These team members have to respect 'the right of the manager to manage and the duty of employees to obey' (Morgan, 1986, p. 187). To some people the unitary ideology may sound naive. However, unitary arguments are often heard and may be extremely powerful. Many theories of management are based on a unitary ideology. An example of how deeply especially unitarism is embedded in organizational life can be illustrated by the concept of 'bureaucratic opposition', for instance 'whistle blowing' (Weinstein, 1979). Whistle blowing refers to an organization member, who makes his or her discontent about some organizational practice publicly known. Even if this particular practice is by commonly accepted standards grossly unethical, bringing it out to the public is hardly ever appreciated by other organization members.

Now that the societal and scientific views underlying research in the social sciences have been uncovered, it is time to proceed to a discussion of the topic 'work design and technology'. However, as will become clear when reading this study, time and again it is necessary to fall back on these societal views, as they underlie much of the literature discussed in the following and partly explain contrasting opinions expressed.

2 Theoretical perspectives

In the previous section, a classification of social theories was presented in which philosophy of science and philosophy of society both played a certain role. A related topic of a philosophical nature concerns the question of the relative importance of theory and practice. Extreme positions in this debate give either one of them priority. A typical comment on theory-driven research, is the assertion, that the research is unrealistic as it has not been tested in 'the real world'. Criticisms on data-driven research include 'inventing the wheel again' and a lack of motivation for the choice of variables. The underlying study consists of a theoretical and an empirical part. The choice for the variables, which are included in the empirical part, is motivated in this chapter.

Most theories, or approaches, discussed in this theoretic chapter are classifiable as sociological, although contributions from industrial engineering, economics and psychology are also used. Studies of a more applied nature concerning the division of labor around automated manufacturing in the Netherlands, are discussed in chapter four.

Doorewaard and Huijgen (1985, p. 56) consider the difference between measures concerning the technical production system and the labor organization as purely analytical. Although changes in the labor organization frequently coincide with changes in the technical production system, the first category may also occur independently of changes in the technical production system. Therefore it is necessary in this theoretical chapter to draw the line between both closely related concepts as sharply as possible. Sections 2.1 and 2.2 deal with the design of, respectively, organizations and technical systems. Section 2.3 discusses one part of the organizational environment, namely the product market, and the consequences this has for an organization's product strategy. Section 2.4 continues the chapter by discussing a second category of environmental influences, namely the labor market. An organization is dependent on the labor market for a sufficient supply of employees with the desired qualities. Two contrasting visions concerning the relationship of demand and supply on the labor market are presented. In the concluding section 2.5, a research model and hypotheses are presented which provide the framework for the remainder of the study.

2.1 Design orientations

Work design involves the assignment of tasks to functions or jobs. Organization design can then be seen as the coupling of all individual functions within an organization. The assignment of tasks to functions can be assumed to follow a specific organizing logic, which is adopted and implemented consciously or unconsciously by organizational decision-makers. This organizing logic is called 'design orientation' in the remainder of this study. In this section, two contrasting idealtypical design orientations are discussed and confronted. Next, attention is paid to Dutch developments with regard to design orientations.

Most discussions concerning the division of labor focus on the design method that is explicitly or implicitly in use to design the production and labor organization. One may distinguish two different design orientations as ideal types. The 'fractionized design orientation', commonly denoted as 'Taylorism', leads to a maximal division of labor. In contrast to the fractionized design orientation, organizations such as 'New Production Concepts' (Kern and Schumann, 1984) are said to be emerging, which are characterized by a minimum division of labor. This orientation will be called 'integrative design orientation'.[1] The adjectives 'fractionized' and 'integrative' cover the most essential characteristic of both design orientations, namely the fundamental principle concerning the assignment of the tasks that have to be carried out in an organization (the 'organizational task pool') to functions c.q. jobs. Thus, they refer to the extent of the division of labor realized. Fractionized designs try to split tasks as much as possible in its constituent elements. Functions are to contain as few tasks as possible. In contrast, in integrative designs functions consist of many tasks, that belong together according to a certain organizing logic. 'Fractionized' and 'integrative' are generic terms or umbrella concepts, covering specific design methods. In addition, they can be used independently from specific periods, industries and countries. Both design orientations have an ideal-typical character, scoring differently on a number of interdependent characteristics, as shown in Table 2.1.

Table 2.1
Design orientations

Characteristic	Fractionized	Integrative
Job content	Small	Large
Hierarchy	Steep	Flat
Organization structure	Complex	Simple
General skill level	Low	High
Basis deployability	Unskilled tasks	Polyvalency

As shown in Table 2.1 the difference in design orientations has important consequences for many organizational characteristics such as job content, the number of hierarchical layers and the number of staff departments, which influence the organization's complexity, the skill levels needed, and the basis for the deployability of employees. Compared with the integrative organization, the fractionized organization is characterized by functions consisting of as little tasks as are necessary to 'fill' a job with respect to its quantitative workload. Many hierarchical and supporting functions exist to control the primary production processes, and to assist production workers. In other words, the division of labor has a horizontal (many hierarchical layers) and a vertical (many supporting functions or departments) dimension. As functions consist of only few tasks, the skills required to fulfil such functions are few. Especially at the lower levèls of such organizations, employees' work is unskilled, which means that employees do not need to possess formal qualifications to fulfil their functions' task(s). This does not mean that such employees may not possess formal qualifications, but only that such qualifications are not needed to perform the tasks of their functions.

The point of 'deployability' is generally called 'labor flexibility'. The concept of flexibility often remains vague, especially if combined with 'labor' as an input in the production processes, and generates much confusion; also because it has a quantitative and a qualitative dimension. Deployability is the extent to which employees are able to perform a variety of tasks. Many tasks require certain skills, which can be learned by employees, but this learning takes a learning-in period. By creating unskilled or 'everybody's tasks' (Kern and Schumann, 1984), fractionized organizations can switch employees from one unskilled position to the other; the learning-in period of unskilled tasks is zero. In integrative organizations, an alternative strategy to take the 'learning-in hurdle' is pursued: employees are polyvalent, i.e. they possess the necessary qualifications to handle a variety of skilled as well as unskilled tasks (Benders, 1990).

The distinction between both design orientations is polar rather than dichotomous. They represent extreme positions on a gliding scale instead of two alternative choices, whereby one particular choice excludes the other.

The brief discussion of both design orientations presented above is elaborated in the remainder of this section, which deals with their historical evolution. The classical way to start such a discussion is to deal with the ideas of Adam Smith. In *An Inquiry into the Nature and Causes of the Wealth of Nations*, Smith mentions the example of pin-making to demonstrate the advantages of dividing labor over many employees (1776, pp. I:7-14). The division of labor leads to an increase of production which would otherwise not be realized. This economic advantage is due to three factors:

1 'improvement of the dexterity of the workman'; 'the division of labour, by reducing every man's business to one simple operation, and by making this operation the sole employment of his life, necessarily increases very much the dexterity of the workman' (1776, p. I:11);
2 'saving of time [...] commonly lost in passing from one sort of work to another' (1776, p. I:12);
3 'application of proper machinery' by 'some one or other of those who are employed in each particular branch of labour should soon find out easier and readier methods of performing their own particular work' or by 'the ingenuity of the makers of the machines, when to make them became the business of a particular trade', respectively, by philosophers (1776, pp. I:13-14).

The classical point to end any discussion about Smith's contribution has just been reached. Lesser known of Smith is the fact, that he explicitly points out, that continuously 'performing a few simple operations' tends to have negative effects on the vividness of the human mind: 'He [...] generally becomes as stupid and ignorant as it is possible for a human creature to become' (1776, p. II:303). Smith considers this as inevitable: 'in every improved and civilized society this is the state into which the labouring poor, that is, the great body of people, must necessarily fall, unless government takes some pains to prevent it' (1776, p. II:303).

Smith was by no means the first to note economic advantages to be achieved by the division of labor. According to Haney (1936, p. 208) the French Physiocrat de Mandeville 'was perhaps the first to use the words "divided" and "division" in' connection with labor. As early as the fourth century B.C. the Greek general Xenophoon made reference of shoemakers being specialized in men and women shoes, which was possible because market size allowed it (Marx, 1867, p. 352). Plato regarded a certain division of labor as 'natural'. Another scholar who discussed the division of labor before Adam Smith is Sir William Petty (1623-1687). Petty describes this phenomenon, without naming it however. He gives examples in which 'each Manufacture will be divided into as many parts as possible, whereby the Work of each Artisan will be simple and easie' so it was possible to produce more cheaply (Routh, 1989, p. 37). Marx (1867, p. 334) mentions several authors before 1750 who explicitly noted the importance of the reduction of labor time for production. Despite these predecessors, Smith is generally accredited for recognizing the importance of the division of labor and systematically describing (some of) its elements. Routh even states: 'Adam Smith takes Petty's idea (without acknowledgement), calls it the division of labour, and makes it the foundation for the growth of wealth' (1989, p. 85).

In 1832, Charles Babbage published the first edition of a book *On the Economy of Machinery and Manufacturing*, in which his experiences were summarized. According to him, the principles stated by previous 'writers on Political Economy' have to be supplemented. He mentions the following principles:

1 The time required for learning depends on the difficulty of the execution of tasks as well as on the number of tasks to be learned.
2 During this learning process, materials are wasted. By restricting the number of tasks to be learned by a person, less waste will occur and the price of production is diminished.
3 The division of labor leads to the saving of time, which otherwise would have gone lost due to changing of occupations.
4 Change of tools leads to a loss of productive time.
5 Skills are acquired by frequent repetition of the same process.
6 The whole attention of an individual performing a task is likely to result in the development of new and improved tools, and, ultimate-

ly, of machines (1835, pp. 169-175).

After quoting Smith's principles as stated above, Babbage points to the 'most important and influential cause [...] altogether unnoticed', namely:

> That the master manufacturer, by dividing the work to be executed into different processes, each requiring different degrees of skill or of force, can purchase exactly the precise quantity of both which is necessary for each process; whereas, if the whole work were executed by one workman, that person must possess sufficient skill to perform the most difficult, and sufficient strength to execute the most laborious into which the art is divided' (1835, pp. 175-176).

The principle of splitting work according to the skill levels needed to perform the different tasks of which the work is composed, has later become known as the 'Babbage-principle'.

Babbage's work was probably a contemporary summary of the state-of-the-art with respect to job design. His views are clearly fractionized. He points to many aspects which are still considered to be important aspects of organizational performance such as set-up times, learning-in periods and the coupling between wages and qualifications. Judging from the fact that his book saw its third reprint within three years, it was probably popular at the time. Furthermore, in his book he makes reference to the Italian writer Gioja, who had published a book *Nuovo Prospetto delle Scienze Economiche* with similar ideas in 1815, of which fact Babbage was unaware until after the completion of his own manuscript. Babbage even thought it 'probable that additional research would enable me to trace of the other principles, which I had thought original, to previous writers' (1835, p. iv). Obviously, similar fractionized design methods were being used and developed simultaneously in different regions. Examples include the Polish-French scholar Skarbek (*Théories des Richesses Sociales*; 1829), the Russian-German economist Storch (*Cours d'Economie Politique*; 1815) and the English writer Ure (*Philosophy of Manufacturers*; 1835).

Another contributor to the debate around the division of labor put the matter into a totally different perspective. Based on an extensive review of literature, both of a theoretical nature as well as reports on the situation of workers in actual factories, Karl Marx formulated an influential societal theory which was echoed both in practice as well as in social sciences (he can be seen the father of radicalism; see section 1.1). In *Das Kapital* a vast number of topics is discussed, among which the division of labor (1867, pp. 323-354). Marx explicitly mentions the economic advantages of the internal division of labor, but having a sympathy for 'workers' instead of 'capitalists', Marx sharply criticizes capitalist production processes ('manufactures'). He observed that many different groups of workers were formed, both skilled and unskilled, and addresses his critique to the presumably widely existing unskilled jobs, although it must be stated that the major element of his work concerns the fact that workers did not own the means of production they were working with. Concerning the division of labor, Marx was to become the major inspirator for the so-called 'labor process approach' (Kern and Schumann, 1970; Braverman, 1974; Friedman, 1977; Noble, 1978; Thompson, 1989), which analyses the division of labor from a Marxist ('critical') point of view.

The name of the American engineer Frederick Winslow Taylor (1856-1915) is the source of the adjective 'Tayloristic' and the noun 'Taylorism'. Taylorism is often used to denote either the fractionized design orientation in general or, be it probably in a minority of cases, the specific fractionized design method developed by Taylor. Taylor's works are clearly unitary, as demonstrated by the following quote from *The Principles of Scientific Management*:

> The principal object of management should be the maximum prosperity for the employer, coupled with the maximum prosperity for each employé. [...] The majority of these men (JB: employers and employees) believe that the fundamental interests of employés and employers are necessarily antagonistic. Scientific management, on the contrary, has for its very foundation the firm conviction that the true interests of the two are one and the same (1911, pp. 9-10).

His book *Shop Management* (1903), of which *The Principles of Scientific Management* was a popularized version, contains similar clearly unitary statements. The basis for Taylor's Scientific Management was his belief that employees, or 'men' in his terms, tend to work slower than possible, which in Taylor's view was either caused by 'the natural instinct and tendency of men to take it easy, which may be called natural soldiering', or by 'more intricate second thought and reasoning caused by their relations with other men, which may be called systematic soldiering' (1903, p. 30). Soldiering lead to lower production and to lower wages than could have been realized otherwise. Scientific Management was directed toward the elimination of soldiering. It consists of four principles

(1903, pp. 64-65):

1 'a large daily task', which was to be 'clearly defined' and 'not easy to accomplish', was to be carried out by 'each man in the establishment';
2 'standard conditions': 'the workman should be given such standardized conditions and appliances as will enable him to accomplish his task with certainty';
3 'high pay for success';
4 'loss on case of failure'.

The basic methods to achieve these principles are scientific time study, standardization of work methods, and the 'differential reward system'. The first principle, scientific time study, is considered to be the most important by Taylor: 'What the writer wishes particularly to emphasize is that this whole system rests upon an accurate and scientific study of unit times, which is by far the most important element in scientific management' (1903, p. 58). 'Each job should be carefully subdivided into its elementary operations, and each of these unit times should receive the most thorough time study' (1903, p. 83). For every elementary operation a 'quickest time[2], in which it can be done by a first-class man' (1903, p. 59), was established by means of time study. The quickest time was used for planning purposes by a planning department and as input for the differential reward system. In this system, a premium was set for the production above a certain level set by time study. Failing to meet this level results in the loss of the premium.

Scientific Management advocated an extensive separation of the planning of work from its execution, which Brödner (1991, p. 1) sees as the new element of Taylorism: 'There is no question that the cost of production is lowered by separating the work of planning and the brain work as much as possible from the manual labour' (1903, p. 121). Work instructions for employees had to be established by a planning department, based on 'scientific' norms. 'Functional foremanship' had to be introduced in which eight foremen were each held responsible for a certain aspect of supervision. This also made the four principles applicable to the level of the foreman. Later, the 'functional foremanship' was to manifest itself in the various staff departments, such as organs for production planning, personnel, quality control and work preparation.

The last element of Scientific Management to be discussed here is Taylor's implicit use of the Babbage-principle. Frequent statements are made to ensure that individual employees should have work fitting their skills: 'the difference between the first-class men and the poor ones is quite as great as that between fine dray horses and donkeys' (1903, p. 189).

Conclusively, Taylor's Scientific Management is functionalist, unitary and individualistic. It is also a bottom-up design method, starting to improve 'shop management' and predominantly limiting the discussion of consequences of the prescribed principles for those parts of the organization that are directly involved in the implementation of these principles. It is a hallmark example of a fractionized design method. Taylor and his many followers provided a more systematic approach, which gave this movement momentum especially in the United States, and later in Europe, but they did not invent the fractionized organization (cfr. van Assen and den Hertog, 1980, p. 52). It is sometimes claimed that Taylor 'was right in the context of his time', i.e. a period when many illiterate workers with a non-industrial background and an insufficient command of English entered American factories *en masse*. After a discussion of Taylor's ideas, Locke even concluded that 'most of his insights are still valid today' (1982, p. 23).

A noteworthy proponent of Scientific Management was Lenin, who advocated its use in the Soviet Union. Before the Russian October revolution of 1917, Lenin condemned Taylorism, but he changed his strand after the Revolution, because labor productivity had to be raised (Lenin 1914 and 1918). Thus, the ownership of the means of production rather than the division of labor was seen as the critical factor.

The merits of Scientific Management as a design method were also, fully or partially, contested by contemporary labor unions (Bloemen, 1988) as well as another prominent management writer, the French engineer Henri Fayol (1916). In contrast to Taylor c.s., Fayol followed a top-down approach. He was concerned with organization design and the application of management principles, rather than the improvement of shopfloor working practices. Fayol formulated fourteen management principles, of which the unity of command was one. On this point, Fayol disagreed with Taylor: 'For every employee whatsoever, an employee should receive orders from one superior only' (1916, p. 31). Taylor's negation of the unity of command is considered to be 'wrong and dangerous' (1916, p. 97), but Fayol agreed with the economic merits of increasing the division of labor (1916, p. 27).

The German scholar Max Weber (1966) advocated the use of 'bureaucratic organization'. Weber's ideal, the 'rational-legal' bureaucracy, is based on rational principles, and exists within a formalized legal framework. Key elements include equal treatment for all employees, reliance on knowledge which is relevant for a certain position, formalized record keeping and standard setting, and binding rules for all organization members, which are also enforced (Perrow, 1986, p. 3). Weber's influence on social theories can hardly be underestimated, but as he did not develop a specific design method, the impact of his thoughts seems to have been indirect. Furthermore, the rational-legal bureaucracy was not a concept invented by Weber, but more the clustering of a number of existing practices in an ideal-typical organization concept.

Before proceeding, three conclusions about the historical development of design orientations concerning the division of labor until approximately the beginning of the second world war can be drawn.

Firstly, the internal division of labor is as old as the phenomenon 'organization' itself. Its origins most likely have to be sought in prehistoric times, and its importance was recognized early in human history. More or less concise methods of prescribing a certain level of the division of labor and forms of organizational structures were developed and implemented. This development took place in several countries, in which organizational and industrial structures were developing within specific societal contexts. Although some exceptions are known, these methods were generally directed toward an increase of the division of labor and can be categorized as fractionized. These methods evolved over time, and there were disputes among their adherents, but their common denominator is their impact on organization and the division of labor in contemporary as well as present organizations. The word 'Tayloristic' does not seem appropriate to describe fractionized design attitudes. Taylorism is only one specific design method out of a broad range of fractionized methods.

Secondly, intellectual thoughts concerning the division of labor have gradually developed from casual observations and descriptions, often concerning a limited number of aspects (Xenophoon), via the realization of its existence (de Mandeville) and first analyses concerning its drives (Smith), to detailed analyses, resulting in more or less precise prescriptive methods (Babbage, Taylor, Fayol). This development not only lead to more detailed methods, these methods seem to have led to increasingly fractionized jobs.

Thirdly, this development was by no means an automatic and smooth process. The process itself was a trial-and-error process, and it was met by (sometimes fierce) resistance from employees and their representatives or sympathizers (Marx). The economic advantages of fractionized organizations throughout history have been such a strong and influential drive that their development was stimulated rather than hindered. Increasing organizational sizes probably formed an important stimulant: economies of scale were to be achieved by larger organizations. The internal division of labor was influenced in at least two ways by the increase of organizational size: more opportunities to increase the division of labor are created, and were used, which called for more complex, generally formalized and standardized, coordination measures.

The diffusion and application of such ideas was and still is a process that proceeds slowly. An example of this is given by Langton (1984), who describes the rise of Wedgwood's earthenware factory in the eighteenth century. During several decades Wedgwood managed to gradually implement several organizational or bureaucratic measures in his factory.

Given this long historical development of fractionized design methods, as well as their economic successes, it is no surprise that their impact still seems to be overwhelming. However, their disadvantages have also been spelled out (Lammers, 1983; Wexley and Yukl, 1984, pp. 278-279; Morgan, 1986, pp. 33-38), which resulted in the development of integrative design methods. Criticisms on the fractionized design orientation are motivated by organizational or social concerns, and can roughly be divided into three partially overlapping categories, namely:

1 Coordination problems within organizations.
2 Organizational inertia.
3 Negative consequences on human attitudes and behavior.

All of these categories are relevant from an organizational point of view, but problems in the last category have also been criticized from a purely social standpoint. Depending on the circumstances, these problems become more or less pressing.

The first category, coordination problems, points to the reverse side of the 'division of labor'- medal. Dividing tasks over many functions automatically implies that the tasks have to be coordinated. Manifestations of coordination problems are manifold. Balancing the capacities of individual work stations in an assembly line and project management are both questions of coordination. Fractionizing labor calls for a high need for coordination. The more persons involved, the more difficult the task of coordination. If many persons are assigned a small part of the total process, it may become hard to see who is responsible for what part, which makes the process intransparent and therefore hard to control. Typical measures to ease coordination are standardization, formalization and bureaucratization. Such measures are often thought necessary when organizations get more complex as they grow.

The second category of problems has to do with organizational change.

Measures taken to facilitate coordination tend to lead to a rigidity, or inability ('inertia') to change. Organization members get used to certain procedures and ways of working, which makes it harder to change. Vested interest may also play a role. The ability to innovate may be negatively affected, which becomes especially pressing when there is a high need for change.

The third category concerns consequences on human attitudes and behavior. Fractionized organizations are characterized by deskilled jobs on lower hierarchical levels. Often, a short training period suffices to perform jobs which are often monotonous. Low commitment, low satisfaction and motivation, the absence of innovative ideas, lack of responsibility, alienation, withdrawal behavior, apathy, carelessness, lack of pride and even theft of organizational properties are just some elements of a long list of (sometimes overlapping) negative consequences of fractionized organizations on their members' attitudes and behavior. Recruitment and motivation problems may result, especially in case of personnel shortages. These attitudes can be seen as organizational disadvantages, but also as negative social consequences per se. In this case, the perspective of analysis is not the organization, but the individual employees within the organization. Objections against these social consequences are often motivated by humanistic concerns rather than by organizational motives. As both employees and organizations may benefit from measures to reduce the division of labor by applying an integrative design method, the achievement of mutual benefits may serve as a powerful argument to use such method. Note, however, that the achievement of mutual benefits has also been used by Taylor, be it that the employee's advantage of Scientific Management was a higher salary or a broader job content in case of integrative design methods. In this way, both Taylor's presentation of Scientific Management and the claimed advantages of integrative design methods are unitary.

The distinction between organizational and social disadvantages is of crucial importance for the development and use of measures to counteract these disadvantages. Humanistic inspired programs such as the German program *Humanisierung des Arbeitslebens* and the French program *Amélioration des Conditions des Travail* can be seen as a reaction to the degrading effect of the presumably dominant fractionized design orientation on blue collar jobs (Lane, 1989). Other integrative design methods emphasize organizational advantages.

Another point that deserves attention is the use of stereotypical portrayals, often by using metaphors (Morgan, 1986), of fractionized and integrative organizations. By using certain images, attention is selectively drawn to some aspects of the organization, thereby simultaneously neglecting other aspects. For instance, fractionized organizations are portrayed either as smoothly operating machines, sharply contrasting with the chaotic functioning of integrative organizations. For both categories, the problems discussed tend to become more pressing under certain circumstances, which will be dealt with further on in this chapter.

Before proceeding to a discussion of Dutch developments in the field of integrative design methods, attention is paid to some influential international developments. While fractionized design orientations are older than Taylorism, 'modern' integrative approaches also date back a considerable period. Although Herzberg et al.'s (1959) influential, yet heavily criticized two-factor theory is sometimes credited for approaches as job rotation, job enlargement, and job enrichment (f.i. van Assen and den Hertog, 1980, p. 57), these terms were used by earlier authors as well. Walker (1950, pp. 54-58) uses 'enriching jobs' and 'job enlargement' to describe experiments at IBM's Endicott plant which started in 1943. The British 'Industrial Fatigue Research Board' published its first report in 1924, in which several forms of job rotation were demonstrated to raise productivity by 20 per cent (Kuipers and van Amelsvoort, 1990, p. 255). The term is also used later by Walker and Guest (1952, p. 148). In the 1930s, experiments with job rotation were conducted in Germany (Kuylaars, 1951, pp. 203-204).[3] These earliest developments found few followers, and are now practically forgotten. Yet, later approaches seem to have been more successful. The first sociotechnical experiments were conducted in the United Kingdom (Trist and Bamforth, 1951), job design was introduced in the United States in 1955 (Davis, 1976, p. 29), and job rotation, enlargement and enrichment came to play a more important role under influence of Maslow, Argyris (1957, pp. 177-185), Drucker (1954, pp. 289-310), and Herzberg. Hackman and Oldham (1980) concentrated on the design of individual jobs, whereas Schumacher's approach is geared to the organization structure. A distinction can be made between methods aimed at individual jobs, and those concerned with the more encompassing organization design. Examples include job enrichment, job enlargement, and task structuring (work design), and sociotechnical design and Schumacher's method (organization design).

The importance of both design orientations differs among countries (Lane, 1989), industries and historical periods. As this study concerns the Netherlands, attention is paid to Dutch developments in the field of design orientations. It will be clear from the preceding remarks, however, that the Netherlands by no means have a monopoly in the development of integrative design methods. As is shown in the following, Dutch developments were influenced by international work.

Dutch developments

The history of 'new' work organizations in the Netherlands is substantial. In their review article, van Hoof and Huiskamp (1989) start the discussion at the beginning of the 1960s, which marks the approximate beginning of more lasting developments. In the following a perhaps somewhat arbitrary distinction is made between sociotechnical theories and other integrative design methods. After a brief description of pre-war experiences, attention is paid to several integrative design methods, namely to work structuring, to a number of experiments which are hard to categorize, and to the development of sociotechnical theories in the Netherlands since 1960.

Bloemen (1988) studied the impact of Scientific Management in the Netherlands from 1900 until 1930. He concludes that the process of splitting tasks was already in full swing before Taylor's ideas began to be implemented. Especially piece-rate reward systems were used which were, next to technical and organizational improvements, a sharper control, and the strict division of labor, reasons for productivity growth. His conclusion fits well in the idea described above, that fractionized designs were already dominant before Taylor's ideas became known. An influential contemporary Dutch author, the engineer van der Waerden, discussed Taylorism extensively (1911). He conducted a survey in many manufacturing industries on the basis of which he concludes, among others, that an increasing division of labor occurs together with continuing mechanization. Few skilled jobs remain, and many unskilled jobs have been created. He explicitly points to the increasing profitability of fractionized organizations. In a later work, he seems to have adopted a fatalistic attitude towards this trend: 'We do not want to embattle the increasing simplification of work [...]; it would be utopian' (1916, p. 72). As human development can no longer be found at the workspot, working time reduction should provide employees ample time to allow for 'spiritual elevation'. Nevertheless, several approaches giving explicit attention to the human factor in production existed in the pre-war period, most notably among them Delft-based engineers (Buitelaar, 1991).

After the second World War, Kuylaars (1951) expresses his concerns about 'mentally drained' labor, the result of applying fractionized design principles. Using data on employment in Dutch manufacturing in 1949, he concludes that 25 per cent of all employees perform tasks which require no schooling at all, while 45 per cent have jobs only requiring a short learning-in period. Only 30 per cent of employees in manufacturing performed skilled labor at the time. Despite this attention to dysfunctional aspects of fractionized designs, the use of Tayloristic principles was still propagated after 1950 by Ydo, a management consultant and founder of one of the larger Dutch consulting firms (Koch, 1988). His plea seems to have been successful. Dutch economic growth was unprecedented in the 1960s. The expanding production capacity was characterized by a rapid mechanization process and 'familiar Tayloristic principles of horizontal and vertical division of labor, standardization of jobs, and so on' (van Hoof and Huiskamp, 1989, p. 156). Shortages on the labor market were solved by recruiting lowly educated immigrants, so-called 'guest laborers', mainly from the Mediterranean area.

In 1960, an experiment was started at a television assembly plant of Philips in Eindhoven (van Beek, 1964; van Berkel, 1968). This experiment was motivated by concerns for organizational issues, and personnel attitudes and behavior. The basic hypothesis was that splitting the long assembly line, which was very sensitive to disturbances, into 'better structured' groups (ultimately) consisting of about 20 persons, positively affected both employee behavior and the organizational problems. 'The organization of labor, and the group size and tasks structure which result from it, proved to be the most important influence on work attitudes' (van Berkel, 1968, pp. 113-114). The positive findings of this experiment had an influence on the development of 'work structuring' within Philips, which is not be underestimated. The term 'work structuring' was introduced in 1963 (Fontijne, 1970, p. 32) and covered a large variety of (piecemeal) changes in the traditional fractionized designs. The early experiments in the beginning of the 1960's 'were not based on a common philosophy or common policy' (den Hertog, 1979, p. 150). About 1965 the importance of work structuring became more officially recognized at the corporate level, job design experiments began to flourish and '[g]radually the practical notion became more influenced by the theoretical notions and ideas. The theoretical framework was eclectic in character; ideas from different schools such as the orthodox job enrichment school, the sociotechnical system approach and organizational development school were included', according to den Hertog (1979, p. 152). By 1968, at least 36 experiments had been carried out. In 1972, work structuring within

Philips was assessed critically by the Philips Central Workers' Council. It was concluded that the movement had scarcely gotten off the ground, only involving 3.5 per cent of Dutch Philips workers, and were often restricted to a small and isolated part of an organization. However, the Council had a favorable opinion about the experiments, and recommended 'preventive work structuring', meaning to apply these ideas to the design of new production systems, plants and technical systems. In a 1977 article, den Hertog mentions some measures to stimulate work structuring after 1972. These measures included central stimulation of work structuring by the concern centre ('social marketing'), an attempt to approach the organization as a whole in contrast to the limited scale of former experiments, an increased worker participation, and preventive work structuring. This last aspect meant to incorporate work structuring into the design of production systems, which is especially relevant in case of heavily mechanized work settings.

Reviewing the Philips experiences with work structuring, Ramondt (1975, pp. 48-76) adopts a sceptical position. External economic conditions, namely a tight labor market and decreasing sales, led to organizational adaptations, including work structuring. The organizational goals were met most of the time, but employees were hardly involved in the reorganizations, which were initiated and controlled by management. Despite the facts that some 54 experiments had been conducted within Philips (den Hertog, 1975, p. 107), and that repeated efforts were made to revitalize and extend the approach, ultimately the movement of work structuring has not taken root (Dankbaar, 1990, p. 14). According to Pot et al., 'work structuring failed due to single sided interest for quality of working life' (1991, pp. 23-24).

Although work structuring did not have a lasting effect on the development of integrative design methods, its value may have been that many experiments were motivated and initiated, which generated experience with a 'new' integrative design orientation. At the time, the distinction between contemporary sociotechnical work and work structuring was vague: de Sitter even called work structuring an 'applied form of sociotechnique' (1974a, p. 65). At that time, the theoretical foundations for Dutch sociotechnical work were constructed. The lack of a thorough theoretical approach in work structuring certainly was an important factor leading to its decline.

Other experiments in the 1960s were motivated, or said to be motivated, by humanistic concerns. Kuipers (1972) mentions 37 cases of 'task structuring', which he uses as a synonym for work structuring. Of these

37 cases, 19 were not situated in the electronics industry, which shows a large overlap with Philips. His data were collected between September 1969 and August 1971, and refer to task structuring and work consultation. Unfortunately, both the description of the data as well as the analysis are superficial. However, it shows that experiments with integrated designs were not limited to Philips.

The protestant employers union VPCW initiated six experiments among its members. These experiments were motivated by 'the thought, that responsibility in its religious/ethical meaning is a factor of sufficient importance to justify experiments with a greater responsibility' (Ramondt, 1968, p. 13), rather than by changing business conditions, as was the case with work structuring. Ultimately, only two of the six organizations involved managed to increase employee responsibility by giving them more control over their own work situation. In one organization, the result was not lasting while in the remaining three cases the reform attempts were stopped within a few months. Overall, the results were disappointing, both from the organizational as from the ideological point of view. 'The process has a sobering effect on those, who expect that idealistic motives - more humane work relations - as such can induce organizational change' (Ramondt, 1974, p. 110).

Ramondt (1974, pp. 76-79) also describes an experiment at the Pernis oil refinery of Chevron. In several departments, namely production, administration and maintenance, more tasks of different natures were added to functions. Management reported favorable results with respect to productivity, organizational transparency and absenteeism. Staff was cut back by 49 per cent. The experiment was said to be motivated by McGregor's theory X and theory Y, but Ramondt dismisses this idea as a social desirable, ex post legitimation of a reorganization, and motivated by economic concerns.

The experiments in the 1960s remained limited to specific parts of organizations and to specific branches. Judged from the point of industrial democracy, the results were limited: 'Only the creation of (semi-)autonomous groups (and to an ever more limited degree, job enrichment) contributed to a greater influence of workers on their own work. However, this autonomy was generally exercised within strict limits. Experiments with autonomous groups, moreover, rather were the exception than the rule' (van Hoof and Huiskamp, 1989, p. 158). The main lesson was that job design should be part of a larger scale process of organizational development, especially with regard to the question how to achieve cooperation for and avoid resistance to the intended changes.

In the 1970s, the buzz words were 'quality of work' (or: 'quality of working life'), and 'humanization of work'. The debate focused more than ever on job content, job requirements and working conditions (van Hoof and Huiskamp, 1989, p. 159). Summarizing projects in this decade, they conclude that management interest in new forms of work organization increased, but remained selective. The approaches used meant to mitigate the effects of organization structures, which were products of fractionized designs. Projects tended to be small-scale, and were generally initiated by management. Huijgen (1984, pp. 13-14) reports on substantial difficulties finding organizations using an active policy with respect to organizational redesign. Out of 300 firms in branches known to make extensive use of unskilled labor, only two had developed a policy the researchers considered to be favorable. Huijgen sees this as a confirmation of the widespread use of fractionized designs in the Netherlands around 1980. However, in this period interesting theoretical, and field research on a limited scale was done in the field of sociotechnical systems design, as the foundations for the Dutch branch of sociotechnical systems design were being laid.

Remarkably enough, the cradle of Dutch sociotechnical experiments was in a service industry (Buitelaar, 1989, p. 75). The first Dutch field experiment was carried out in 1965 in an organization, clearing large numbers of cheques (van Beinum, van Gils and Verhagen, 1968). Table 2.2 summarizes the basic approach.

Table 2.2
Summary of field experiment in cheque clearing department

Old system	New system
Simple individual tasks	Complex task groups
Complex organization	Simple organization
External control	Internal control

Source: van Beinum, van Gils and Verhagen, 1968, p 46

The experiment was influenced by sociotechnical theories, and conducted with involvement of the Tavistock Institute. An important reason for conducting the experiment was the tight labor market. Experiments in the same organization had been carried out by van Beinum (1963), but these experiments hardly influenced the organization of labor. 'Joint optimization' of the technical and social system were assumed to lead to the most effective organization. The old organization was characterized by individual, independent and homogeneous tasks, and task specialization. Employees were not able to take over each other's work, and close supervision was felt to be necessary. Four changes were implemented: group tasks, task variation, promoting the knowledge about work coherence, and giving more responsibility.

The results were that job satisfaction, work morale, and productivity increased. It remains unclear whether the rising productivity can be attributed to the experiment, as only one (out of two) of the control groups showed a similar development.

Allegro's study (1973) in a textile mill is influenced by sociotechnical work and by organizational development. After discussing the content of sociotechnical reforms, Allegro criticizes, among other things, the neglect of the process of organizational change of the contemporary sociotechnical systems theory. After two-and-a-half years, 40 employees performed '[s]mall group tasks for 2 persons [...] which had the characteristics of an autonomous group task.' Productivity had not decreased, employee involvement had increased, and absenteeism had decreased. These changes were attributed to new personnel and to the new organizational policy. 'However, the most important result seemed to be that the organization is actively involved in a learning process' (Allegro, 1973, p. 210).

Later, Allegro and his co-author Kanters (1975) again stressed the importance of the change process rather than structural variables. In laboratory experiments, they found little evidence for the existence of a (positive) relationship between the perceived degree of enrichment and intrinsic work satisfaction, involvement, and quality of working life. Indications for such a relationship were only revealed if process variables were taken into account.

Besides some more scattered field projects, important theoretical work was done by de Sitter and his associates at Eindhoven University of Technology (de Sitter 1974a and 1974b; Van der Zwaan et al., 1974; van der Zwaan, 1975). The lack of a sound theoretical foundation was felt to hinder the development of sociotechnical work, and was seen as an explanation for the stagnation of the sociotechnical developments at the British Tavistock Institute. Thus, a theoretical, and therefore transferable, systematic approach had to be developed. At the time, de Sitter called work structuring 'an applied form of sociotechnology' (1974a, p. 65), which is logical given the contemporary lack of a theoretical foundation. *Sociotechniek* is defined as:

> the study and explanation of the manner in which both the technical instrumentation and the division of labor are related to environmental conditions and the operation, capacity and output of production systems, as well as the application of this analytical insight into the design of production systems (de Sitter, 1974a, p. 81).

Important items in this early theoretical work include the emphasis on the interconnectedness of the production process, division of labor and the technical instrumentation. The production system is aimed at producing output, for which norms exist. It functions in a dynamic environment, which may lead to organizational changes. This has consequences for process regulation. The control capacity to regulate a process is a vital element in Dutch sociotechnical thinking. The control structure has to be a part of the system it aims to control.

The same reasoning is also applied to an important system element, namely job content. Employees need to be able to solve deviations of standard situations, such as disturbances, independently. To do this, employees need to possess control capacity, i.e. controlling tasks necessary to perform the executing tasks assigned to them (cfr. 'job enrichment'). Later, Karasek's study (1979) on the relationship between job decision latitude and job demands on the one hand, and employee well-being on the other hand, confirmed the relationship between objective job characteristics, and the employee's health complaints c.q. stress indicators (higher job decision latitude in relationship with higher job demands correlate negatively with the occurrence of health complaints). De Sitter's hallmark study 'On the Road toward New Factories and Offices' (1981) brought the approach to the attention of a large public, and can be seen as a catalyst for the development of Dutch sociotechnical thinking. In it, extensive use was made of Swedish experiences.

Terms such as 'sociotechnical systems design' and 'sociotechnology' cover a broad range of approaches. Van Eijnatten (1991) even distinguishes four different generations of sociotechnical thinking. Much of the work outside the Netherlands stresses the participation of employees in systems design, whereas most Dutch work emphasizes the use of specified design rules, and can be labled as 'an expert approach'. Currently, the most important design method is 'modern sociotechnology', MST (de Sitter, 1989). Its importance is also acknowledged by a former foreman of work structuring. Reviewing a report about a 'radical application of sociotechnical systems by design' by Digital Equipment in Scotland, den Hertog notes, that the case 'adds little to our knowledge of the structural side of organization renewal. [...] one has to start with the redefinition of the production structure' (1991, p. 143). Obviously, the former 'eclectic' approach has been abandoned by den Hertog in favor of MST.

Although the approach is quite complex and several more or less detailed variants are applied (Buyse, 1987; van Eijnatten, 1985; Groep Sociotechniek, 1986; Kuipers and van Amelsvoort, 1990; Pot et al., 1991), an attempt is made to sketch its major principles, based on an summarizing publication by de Sitter (1989; also see de Sitter and den Hertog, 1990). MST involves a number of steps, aimed at certain design parameters. Although attention to employee participation is present, these prescribed steps need to be followed more or less strictly. The first step is 'parallellization of the production structure', which refers to the clustering of products into groups. The aim is to reduce the variety of product flows by creating independent parallel flows preferably corresponding to market combinations. The design process of MST thus starts top-down.

The second step involves the 'segmentation of the production structure'. The operations necessary to make the product groups are clustered into segments, based on the reduction of interfaces between the segments. By also assigning all supportive and preparatory tasks, such as planning, maintenance and planning, the number of interfaces is further reduced, and the segments can function (relatively) independently from other segments. In the third step, so called 'whole task-groups' are formed, which are to carry out all tasks within a certain segment. In the ideal situation, all group members ought to be capable of fulfilling all possibly occurring tasks.

The next step involves the design of the control structure. The reduction of the number of interfaces in the previous steps leads to a reduced need for control, as it is no longer necessary to manage a substantial number of interfaces. The remaining control functions are assigned to the whole task-groups, which enables these groups to handle problems themselves, as far as such problems can be solved without consequences for other groups. The control structure is designed bottom-up, assigning control tasks as low as possible in the organization. The same approach is used for the design of individual jobs. MST considers job content as the most important element of the 'quality of working life', although job content is acknowledged to have consequences for the internal labor relations. Following the system theoretical framework, a job ought to consist of executing tasks as well as those controlling tasks, that are relevant for the execution of the executing tasks. The individual must have sufficient job decision latitude to make decisions concerning his own work process. Furthermore, there must be sufficient variation in the task structure of jobs. The approach is considered to be 'objective', i.e. high quality jobs can be designed following design rules. The job holder's 'subjective' view is not taken into consideration.

Another element that merits attention in MST, are the differences between MST and earlier sociotechnical theories. An important claim of MST is, that it is aimed at an 'integral design', which means taking all aspects which are relevant for the new organization design into account simultaneously, instead of focusing on one certain aspect. Rather than focusing on the improvement of jobs characterized by low work content, MST sees the traditional fractionized organization as the root of problems with respect to organizational functioning as well as work content. Although the approach may have been and still is advocated in the first place because of the resulting improved jobs (Pot et al., 1991), its goal is broader than just the 'quality of working life'. MST's goal is organization, of which job design is a part.

In the Netherlands, MST is flourishing, at least at academic and research institutions. Many consulting firms make use of MST as well, and several publications of the Labour Inspectorate, which is a part of the Ministry of Social Affairs and Employment, stimulate its use (Projectgroep WEBA, 1989; Pot et al., 1991; Peeters and Pot, 1991). Nevertheless, a multitude of hiatuses is felt to exist. The MST approach has hardly paid attention to supportive functions such as personnel management and accounting, and specific requirements with regard to suitable technical systems in the fields of manufacturing, production planning and control, and information systems. The process of organizational change is another area which needs more attention. The 'objective' approach to job content is not uncontested, and has been supplemented by adding additional criteria (Projectgroep WEBA, 1989).

As with other design methods, the impact of MST on Dutch design practice remains unclear. The rising number of cases using this method wholly or partially, which are reported in the literature (Kuipers and van Amelsvoort, 1990) seems to point to its increasing acceptance, predominantly in manufacturing industries. The effects of 'autonomous task groups' in thirteen organizations have been studied (COB\SER, 1990). Nine of these task groups are situated in manufacturing industries. According to den Hertog, 'it proved to be extremely difficult to find a sufficient number of innovating companies' (1990, p. 126) for this study. More precisely, almost one hundred organizations were approached. In 'many cases', there was no experience with task groups. The COB\SER-study is in its terminology clearly influenced by MST, but fails to distinguish between the application of task groups as a result of MST and other uses of task groups, although in all thirteen cases a 'flow oriented production structure' was implemented. Furthermore, job enlargement, job rotation and job enrichment are reported to have occurred, but the effects of these different forms are not reported. These are serious flaws as far as the specific effects of MST are concerned. However, considering any implementation of task groups as a move towards integrative designs, the study does provide some interesting results.

In the first place, in eleven cases the reorganization was motivated by economic concerns such as lowering unit costs, improving quality, and shortening delivery times. There were two cases in which 'issues of personnel and organization, namely 'reduction of absenteeism and improvement of the quality of working life', were prime motivators. However, a drive to reduce absenteeism is probably motivated by economic rather than humanistic concerns. Furthermore, it was not even realized in all organizations that the introduction of task groups would have consequences for the quality of working life. Given these findings, little doubt can exist about the importance of economic motivations for these moves towards integrative organizations. Despite this, the quality of working life improved in all but one case. This was measured by taking into account criteria with respect to the 'completeness' of a job, the presence of regulating tasks, possibilities to make contacts, autonomy and complexity. Furthermore, almost no employee expressed a desire to return to the previous situation. These results fit well with the sociotechnical claim, that the extreme division of labor is the root of problems with the quality of working life as well as organizational functioning.

Conclusions and discussion

Organization and job design are core items for economic performance. Both fractionized and integrative organization designs have been attributed superior qualities, be it in different periods and by different authors. Where the economic superiority of fractionized designs went uncriticized for centuries, they have been under attack for several decades now. Despite the negative features ascribed to them, the fractionized design still seems to dominate organization and job design. If so, the arguments for an integrative design have not (yet?) been sufficiently convincing, or put differently: integrative organizations have not been able to economically outperform fractionized ones. Repeatedly, integrative experiments were cancelled, which does not make sense from an economic point of view if they really showed a favorable economic performance. This leads one to adopt a sceptical attitude toward the claimed economic advantages of the integrative design. Stubborn managerial conservatism to stick to fractionized designs is not a sufficient explanation for their persistence, if this conservatism lacks an economic foundation.

As shown above, the field of organization and job design suffers from a lack of historical reflection. Given the close ties between the success of a specific design method, and the monetary and non-monetary interests of their proponents, it is easily understandable that the field is vulnerable to fashions, whereby a specific design method is claimed to be unique and new. As a science, however, organization and job design would be helped by an inventarization of the advantages and disadvantages of both designs as objectively as possible. Such inventarization will enable a more conscious choice for a specific design. Undoubtedly, it will be severely hampered by the fact that integrative designs are desirable from a humanistic point of view. However, humanistic concerns seldom prove to be sufficient for the implementation of integrative designs: they have to be supplemented by economic motives. In this respect, the proven economic advantages of the fractionized design, which probably are recognized in practice, also have to be acknowledged. Combined with the substantial difficulties in demonstrating the economic benefits of integrative designs (Hopwood, 1979), this may be a further obstacle for integrative designs. On the other hand, changing environmental conditions may stimulate the adoption of integrative designs, as will be discussed in the sections 2.3 and 2.4.

But this verdict may be too harsh. Based on large-scale research in the German chemical, machine-building, and automotive industry, Kern and Schumann signalled a tendency toward more integrated organizations in these branches in 1984, prophetically called *The End of the Division of Labor?*. In follow-up research, this tendency is confirmed (see the post scriptum of the 1990 edition of their book and Schumann et al., 1990). Yet, when Dankbaar discussed Kern and Schumann's work together with other integrative scenario's such as Piore and Sabel's and de Sitter's, he adopted a critical position. These authors imply a reversal of the trend toward an increasing division of labor, as it is in the interest of management. But they are 'too much future-oriented, too speculative with too little empirical support for their positions, and, finally, too optimistic' (Dankbaar, 1988, p. 30). Furthermore, descriptive and prescriptive opinions mingle, and the generalizability of the suggested trend to other branches and countries is questionable at best. Research on a broader, preferably statistical and longitudinal scale is needed to find out whether or not fractionized organization designs are abandoned in favor of integrative ones. Schumann et al. report on such follow-up research (1990).

On the other hand, the refinement of the fractionized design has taken several centuries at least. Although especially social disadvantages of the fractionized design were signalled at least as early as 1776, doubts about their economic superiority date back not more than 70 years. As happened with fractionized design methods, integrative designs are in the process of slowly being refined and improved. It can not be expected that they are implemented on a large scale overnight, and ultimately, they ought to be based on a solid theoretical and empirically validated approach. The Dutch post-war attention for integrative designs, with MST as the current state-of-the-art, is just one illustration of the arduous work toward 'the end of the division of labor'.

2.2 Technical systems

Numerous studies concerning the influence of the introduction of a specific technology on the division of labor have been, are, and will be conducted. As the underlying study belongs to this category of studies, attention will have to paid to 'technology' in general. The central question in this section is to what extent technology influences the division of labor within organizations. An answer to this question is given by discussing various notions of the concept 'technological determinism'.

A popular way for opponents of technological determinism to start their discussions is by elaborating on this concept, as in the following example:

> In order to form a new understanding of the relationship of technology to organization and job structure, the existing dominant appreciation needs to be made explicit. The present view, carefully nurtured for the last 150 years, is that of technological determinism, and it is dangerously simplistic. [...] Technological determinism maintains that technology evolves according to its own internally derived logic and needs, quite independent of social environment and culture. Further, it holds that to use technology effectively and thus secure its benefits for society, its development and application must not be inhibited by any considerations other than those determined as relevant by its developers - the engineers or technologists. [...] The claim is made that organiz-

ational structure and behavior is predetermined by technology and is unalterably locked into its needs (Davis and Taylor, 1976, pp. 379-380).

The idea behind technological determinism is clear: technology does and should prescribe the division of labor within organizations. Its development should not be hindered by non-technical considerations.[4] Davis and Taylor seem to see such technological determinism as an adequate description of the status quo (in 1976) and it also has a prescriptive character. However, it is rather difficult to find outspoken representatives of technological determinism. Technological determinism may be a concept used by social scientists to denote a technical preoccupation which is often present in designing manufacturing systems. As this task is often left to mechanical engineers, who are generally (kept) unaware of social design criteria, a technical focus with little or no attention for human aspects remains. This does not hold true for ergonomics or human factors engineering (Perrow, 1983), but this approach is a far cry from mechanical engineering. Furthermore, ergonomics seem to be dominated by an attitude to adapt physical working situations to human requirements rather than to change job designs.

Another interpretation might be that other social scientists are considered to be technological determinists (which is an effective strategy to make the accused's writings doubtful). Jackson, Morgan and Paolillo (1986, p. 244): 'The origin of the view that technology determines organization is usually assigned to a study by Joan Woodward of manufacturing firms in southeast Essex, England, although Woodward herself did not claim a causal relationship'. Examples of this practice can be found in Batenburg (1991, pp. 15-16), Kieser (1988, p. 321), Koopman-Iwema (1986, p. 5) and Miller (1988, p. 336). Woodward herself does not share the view of her being a technological determinist: 'It is not suggested that the research proved technology to be the only important variable in determining organizational structure' (1965, p. 50) nor that 'the research produced anything approaching a general law about the relationship of technology to organizational behaviour', but 'it did suggest, however, that technology [...] is casually related to the structural and behaviour variations observed in manufacturing situations' and 'Organizational structure is the complete product of an unknown number of variables' (1965, p. 248). Although these quotations can hardly be understood or interpreted as technological determinism, the relationships between organization structure and technology observed by Woodward, were often explained as such. Perrow foresaw this danger. Reviewing

Woodward's 1965 book he warned: 'the theory can be construed as technological determinism' (1967b, p. 313).

Before the question whether technological determinism exists can be answered, the concept of 'technology' itself needs to be clarified. Child states: 'The term technology is employed in almost as many different senses as there are writers on the subject' (1974, p. 14). Fry (1982) published an review article of empirical studies on technology, carried out between 1965 and 1980. In his article, six different ways to operationalize technology were distinguished:

1 Technical complexity (Woodward, 1965).
2 Operations technology (Aston Group).
3 Operations variability (Aston Group).
4 Interdependence (Thompson).
5 Routine - nonroutine (Perrow, 1967a).
6 Manageability of raw materials (Mohr).

His conclusion is not surprising: 'Overall, researchers have conceptualized technology differently in terms of both its meaning and dimensionality' (1982, p. 541). Huijgen (1984, pp. 64-65) adds to Fry's list some other interpretations of technology, all stemming from organizational sociology. He makes some important observations about the concept of technology as used in organizational sociology.

Firstly, some interpretations contain elements of the organizational structure, which makes it difficult to draw a clear line between technology and characteristics of organizations. Similar contaminations are found in standard textbooks (see for example: Kast and Rosenzweig, 1981, p. 176; Robbins, 1984, p. 175). It follows from these conceptions that technological determinism in the sense of an overwhelming influence of technology on the division of labor and organization structure can not exist without falling into circular reasoning. As (elements of) the division of labor, the organization structure itself and\or qualifications of organization members on the one hand and technical systems on the other hand are considered to be part of the concept of technology, it makes little sense to investigate the influence of technology on some element(-s) of technology.

Secondly, the interpretations and operationalizations of technology tend to be very abstract, which may be inevitable when different types of organization, such as schools and hospitals, are studied and compared. The often contrasting results of such studies can in many cases be attributed to such 'semantic' problems, as Hunt (1971, p. 108) calls them.

For the scope of this study, the concept of technical systems as used in industrial sociology is more relevant than the broad conceptions, and therefor unnecessarily broad definitions, used organizational sociology (cfr. Lammers, 1983, pp. 80-90 and Aldrich, 1972). Thus, 'technical systems' and 'techniques' will be used, which are more concrete concepts than technology. Given the above discussion, however, it is surprising how many studies on the influence of a specific technical system on some aspect(-s) of a labor situation carry the term 'technology' in their title. Hunt (1971) also chose the relatively small concept of technical system. According to Huijgen (1984, p. 65), this is also the common interpretation of technology in popular views. 'The technical system can be defined as the collective instruments introduced into the action system, including all physical influences which operate in conjunction with members but which are themselves systems independent of members' (Hunt, 1971, p. 100). These technical systems may require humans to be operated, but their way of operation is determined by the technical systems themselves. These technical systems can be analyzed in three ways:

1 the degree of choice whether or not to use the technical system;
2 the technical complexity of the instruments including the number of instruments in the process;
3 the complexity of the technology involved. In Hunt's opinion, technology includes energy, tools and machines, materials and inputs, transportation and communication as well as (organization) members' skills and knowledge (this notion of technology fits well in the broad conceptions of technology which were mentioned previously).

Technical systems are assumed to have an important impact on the factor 'labor'. To investigate this relationship it is necessary to use these concepts in a way that any contamination is eliminated. This is possible by using technical systems rather than technology.

If 'technical system' is the common interpretation of 'technology', one wonders whether or not some form of technical determinism exists instead of technological determinism. Put differently: to what extent does a specific technical system determine the division of labor?. The key to the answer is given by Hunt: 'the instruments may very well need members to operate them, but the method of operation is determined by the instrument itself' (1971, p. 100). In other words, once a technical system is installed, the tasks that have to be performed in order to keep the system running, are determined. As the division of labor involves more variables than only the task pool around technical systems, the technical system is just one out of a number of variables determining the division of labor.

This partial technical determinism is a valid reason to study the influence of technical systems on the division of labor as long as it is realized that at the same time many other factors also play a role, which is at least equally important.

As technical systems determine the task pool around themselves, attention has to be paid to the design of these systems. By influencing the design, the task pool can also be influenced: 'if certain human choices guided the course of its development in the past, other choices could steer it in the future' (Noble, 1978, p. 346).

Design orientations and technical systems

In a report 'Strategic Options for "New Production Systems" - CHIM: Computer and Human Integrated Manufacturing', edited by Brödner (1987), a distinction is being made between two different ways of designing technical systems, namely the technocentric and the anthropocentric way. Although the term 'anthropocentric' is used instead of 'human centred', the contribution of Corbett (1987, pp. 157-173) in this report clarifies best what the difference is between what can be considered as technical design orientations.

Corbett argues that the historical trend in the design of shopfloor production work has been toward the incorporation of human skills into automatic machinery, which he calls 'human-machine comparability'. Operators are left to perform those tasks, which can not yet be automated economically. Typically, these are monitoring and error recovery tasks. In this view, the human element of a production system is regarded as a source of uncertainty, that ideally has to be eliminated. At the same time, however, the human element has to cope with unforeseen events, which makes system performance dependent upon the human factor. Despite that, the typical technocentric way to deal with unforeseen disturbances is to try to predict all possible disturbances, and develop hardware and\or software to control them. According to Corbett, the legacy of this technocentric approach is as follows:

> Having fragmented the job of the [...] operator and removed a large part of his/her skilled control of the system, the designer then relies on him or her to deal with the unforeseen (unprogrammed) disturbances. However, a technical system that does not provide the experience out of which operating skills can develop will be vulnerable in those circumstances where human intervention becomes necessary (Corbett, 1987, p. 161).

Following Bainbridge (1983), Corbett calls this one of the 'ironies of automation', but 'independency paradox' may cover the phenomenon better. This independency paradox can be dissolved by adopting the anthropocentric idea of 'human-machine complementarity', i.e. humans and machines both have their own advantages and disadvantages, which can complement each other. Machines can process and analyze (standard) data quickly, while humans are strong in pattern recognition, synthesis and intuitive reasoning.

An additional problem with technocentric designs is that they often require skills, that are unrelated to existing skills, which leads to training and utilization problems. Anthropocentric designs have to take skill requirements into account, which is possible as choices can be made regarding the allocation of tasks to the technical and human subsystems. Furthermore, attention has to be paid to control and informational characteristics of the human-machine interface, which is a typical subject of ergonomics.

Table 2.3 summarizes some differences between both technical design orientations.

Table 2.3
Technical design orientations

	Technocentrism	**Anthropocentrism**
Allocation of functions	Elimination of humans Comparability Rigid	Utilization of human skills Complementarity Flexible
Control characteristics human-machine interface	Computer in control by pre-programmed actions Humans to monitor	Human in control through human skills Decision support by computer

Source: Adapted from Corbett, 1987, p 167

Examples of human centred systems include the Scandinavian UTOPIA project in the graphic industry, and the UMIST project on human-centred software for a CNC-lathe, which was later funded by the European

ESPRIT program (Corbett, 1987, pp. 169-170), and amounted in a larger scale project also involving computer-aided design and production planning (Murphy, 1989; Gottschalch, 1990).

The use of human centred systems involves more than just user-friendliness. However, user-friendliness can be an important aspect of human-centred systems. According to Brödner (1991, pp. 10-15), the traditional abstract programming methods and editing devices are not compatible with the way skilled workers think and work. Therefore, some alternative programming methods (*WOP* = *Werkstattorientiertes Programmierverfahren*; 'Shopfloor Oriented Programming') have been developed, which allow skilled workers to use their situational and manufacturing knowledge, as well as their planning capabilities. Typical elements of these methods are a database of preprogrammed forms such as geometric modules and cutting cycles; programs can be written by combining modules stored in the database, and setting parameters. Other elements include a tool database and graphic, real-time simulation. Brödner states that workers with no previous programming experience were able to deal with this way of programming after a five-day learning period. This example is important in that it points to the fact that diverging skill requirements may be a reason to opt for fractionized jobs, i.e. a separate machine operator and programmer. However, bringing programming skills in line with the skills possessed by traditional workers is by no means a guarantee for the occurrence of integrated jobs.

Analyzing the difference in freedom of choice between a stand-alone machine and the design of larger production systems such as assembly lines, one will find that the extent to which the technical design influences choices with respect to the division of labor varies. Compared with stand-alone equipment, once designed (and constructed) more complex (measured by the number of elements, and their interconnections) production systems, allow less degrees of freedom, especially if work stations occupied by humans are included in the design of the production system. It is likely that complex production systems are not bought 'off-the-shelf', but are designed by project groups mainly consisting of engineers. In the design phase, critical choices regarding the production structure and division of labor are made, and incorporated into the production system. This realization underlies the plea of van der Does de Willebois (in: Ramondt, 1974, p. 60) to take social and human criteria into account with the design of production systems, and the notion of 'preventive work structuring' (den Hertog, 1979), as well as the plea to incorporate criteria with respect to the functioning of humans in the design of production systems (Brödner, 1991; Kidd, 1991). With this plea, the possibility to take such criteria into account in the design process is recognized.

The design of stand-alone equipment is another matter. Stand-alone equipment is generally bought off-the-shelf. 'Tailor-made' machines exist, but may be too expensive to be bought by most organizations, especially small and medium sized enterprises. For many buyers of machines, the possibilities to influence the technical design are small, or even non-existent. On the other hand, stand-alone machines generally leave substantial room for taking decisions with respect to the organization of labor around the machine. Whereas workspots in a large assembly line are incorporated in the technical design, and hardly any opportunity is left to change the line once implemented, a stand-alone machine such as a ribbon-saw only determines the way it has to be operated, i.e. what tasks have to be conducted in order to have the particular machine function. The division of these tasks is in no way determined by a stand-alone machine.

This difference has implications for the relevance of aspects of both technical design orientations. In the design of complete production systems in which the division of (human) labor is already to an important degree incorporated, aspects relating to the division of labor such as machine-paced jobs and the inclusion of discretionary tasks into jobs, are also critical. For stand-alone machines, user-friendly software and ergonomic aspects are more important, but the division of labor is in the first place a matter of the organizational design orientation used. As Murphy states: 'the human-centred technology must be integrated with, and complemented by, an equally flexible organisation of work' (1989, p. 148).

Having discussed organizational and technical design orientations, their relationships can be dealt with. In the following, an attempt is made to answer the question whether or not this relationship can be considered to be neutral, in the sense that any organizational design orientation can be combined with any technical design orientation. If so, this fits in the idea that degrees of freedom exist around technical systems, and that technical determinism has to be rejected. If not, it has to be possible to identify technical characteristics of technical systems suited to be used together with a certain design orientation.

		Organizational design orientation	
		Fractionized	Integrative
Technical design orientation	Techno-centric	I	II
	Human centred	IV	III

Figure 2.1: Technical and organizational design orientations

Certain combinations of design orientations and technical systems, as designated by the quadrants of Figure 2.1, are more or less likely to occur. There are several possibilities:

1 Total freedom of choice: All combinations are equally likely to occur, which represents the situation of total freedom of choice.
2 Technical determinism: Quadrants I and III are the only possible combinations, i.e. technocentric systems lead to a fractionized design, and human centred systems lead to an integrative design.
3 Intermediate positions.

Ad 1: Total freedom of choice If the freedom of dividing tasks around a given technical system is total, all four quadrants of Figure 2.1 present equally likely options. Technical characteristics of technical systems do not correlate with the design orientation in use. If this position is accepted, it hardly makes sense to make a distinction between technocentric and human centred systems. The call for 'computer and human integrated manufacturing' can then be understood as a warning not to focus on one aspect single-mindedly, but to take both the technical system and the design orientation simultaneously into consideration. Such a plea is well known in the (early) sociotechnical call for 'joint optimization' (van Eijnatten, 1991, p. 11, attributes this concept to Emery) and the concept of 'synchronous innovation' (Bennett, 1992). But as many authors stress the importance of developing human centred systems, and as numerous studies on 'technology and work' exist, one may assume that technical systems do have an important influence on jobs, and are not neutral.

ad 2: Technical determinism Another possibility is that technocentric systems fit with fractionized designs, and that human centred systems fit with integrative designs. In that case, only quadrants I and III are possible. In this case, technical systems possess characteristics that can only be complemented with a particular design approach. An organizational design orientation may even be incorporated into technical systems. However, as with technological determinism, it is impossible to find authors defending such a radical standpoint.

ad 3: Intermediate positions Not surprisingly, many intermediate positions can be taken between technical determinism and total freedom of choice. In the following, two such positions are discussed, namely technical prescription and technical seduction.

Braczyk (1987) provides a moderate example of technical prescription. Based on research in the German clothing industry, he distinguishes between two contrasting types of technical systems in clothing production:

> The first type by design permits both the use in a rigid Taylorian assembling line and in a restructured work organization consisting of different semi-autonomous working groups which perform at relatively high level of skill. In this case the real use depends on the choice the user is taking on how to use the system.
> The second type is strictly designed according to the principles of a Taylorian assembling line without admitting any alternative (1987, p. 149).

In terms of Figure 2.1, only quadrants I, III and IV are relevant in Braczyk's case. It has be noted that this involves the design of a production system, namely an assembly line in which work stations are included. As Braczyk's second type shows, once this design is established, a fractionized organization of labor is prescribed by the technical system. However, an integrative organization design can not be prescribed. As is evident from the first type, the technical design may permit choices concerning the organization of labor, but there is no guarantee that an integrative design is chosen. In this case the choice for an integrative design is not obvious, but optional. Note that this case of partial technical prescription involves a production system, not a stand-alone machine.

Labor process theorist Noble (1978) also attributes a critical role for technical systems. Noble argues that one important reason to adopt the Numerical Control (NC) - technique in favor of the alternative Record

Playback system was the fact that in contrast to Record Playback, NC provides the possibility to split controlling and executing tasks, and thus fits well in a fractionized design orientation. In contrast to RPB, NC provided the opportunity to formally synthesize machinist skills, and transfer them to a central programming office. Noble stresses that this transfer is not necessary, but is made possible by NC (1978, p. 340).

Whereas Noble stresses the use of NC as an instrument in a fractionized design orientation, the choice for a fractionized design may not be as conscious as he asserts. The choice for fractionized jobs may just become more likely because a task such as programming poses totally different, namely conceptual, skill requirements than machine operation, which requires manual skills and machining experience. This gap between the skill requirements of programming respectively operating may induce the application of fractionized design principles, especially in the period and country studied by Noble. As Davis, Canter and Hoffman's (1955) empirical study indicates, the fractionized design orientation was dominant in the USA during the 1950s, which makes it likely that fractionized jobs were commonly found in the US industry in general as well as around NC-machines as a specific example. NC merely made these specific forms of dividing labor possible.

In this respect, Manders' notion of 'partial rationality' (1990) is interesting. Manders studied the process of decision-making at the 'Centre for Manufacturing Technology' at Philips Eindhoven, an engineering department developing systems used by Philips. Manders concludes that control of labor is not a conscious goal of top management and R&D-management. However, Tayloristic concepts were transformed into technical concepts. These have become part of the curricula of technical schools, and the education of (mainly mechanical) engineers. Via socialization processes, both in formal schooling and in everyday working practice, Tayloristic concepts have become widespread in the field of engineering, and have become 'the way to do things'. The validity of such designs is never questioned just because of their widespread, albeit implicit, acceptance, which is typical for the partial rationality of engineers (cfr. the notion of 'paradigms' in science).

Regarding technical prescriptions, it can be noted that the complexity of technical systems influences the degree to which choices regarding the organization of labor can be made. Braczyk shows that large, complex production systems may already to an important degree incorporate the organization of labor, whereas Noble, focusing on stand-alone machines, shows that NC-programming allows different choices, but can be used as an instrument in a fractionized design. Finally and in contrast to Noble,

Manders argues that such a use may not be deliberate but can be a result of the partial rationality of designers of (technical) production systems.

The notion of technical seduction can perhaps best be illustrated by the example of production planning and control (PPC) systems. According to Peters and Kommers (1991, pp. 48-55), many organizations buy standard software packages for production planning and control purposes. Although it is not necessary to use all modules of such large, standardized packages, the temptation to do so is great. The criticism of technocentric PPC-systems is that shopfloor planning is centralized, thereby limiting the autonomy and flexibility on the shopfloor.

Manske (1987) introduces two alternative strategies to use PPC-systems (cfr. Nullmeier and Rödiger, 1986), namely 'centralized total planning' and 'framework planning'. Centralized total planning tries to achieve complete control of production, including planning of demand and production at firm level, material requirements planning, capacity planning, inventory control, and shopfloor planning (see Vollmann, Berry and Whybark, 1984, for an example of this approach). They can be seen as representing the Tayloristic ideal of complete management control. Manske's main criticism of centralized total planning is directed at the detailed shopfloor planning prescribed. This leaves little freedom for shopfloor workers, and production plans are hard to adapt to changing circumstances (for instance due to rush orders, machine disturbances and planning errors). These last aspects limit the applicability of total centralized planning. The alternative framework planning leaves shopfloor planning of orders to shopfloor workers.

Although it is possible to write tailor-made production planning and control systems, and centralized planning software can be adapted to the control structure of production organizations, organizations may be seduced by the technical possibility to plan shopfloor activities in detail. In other words: although a centralized production planning package does not necessarily lead to fractionized organizations, this combination of quadrant I becomes likely. The ultimate choice, however, is an organizational choice (Alders, Fruytier and Peters, 1991; Peters and Kommers, 1991).

Conclusion and discussion

Having discussed various positions with respect to the question to what degree technical systems influence the division of labor, it can only be concluded that this degree varies, and is influenced by a number of factors, such as:

- the complexity of technical systems (stand-alone machines vs. complete production systems);
- the degree to which technical systems are tailor-made;
- the moment design decisions are taken, combined with the degree in which such designs are influenced by users;
- engineers' awareness of human and social design criteria.

Technical systems, especially large production systems, may implicitly or explicitly reflect a design orientation. Whether or not technical system designers are aware of and take into account criteria with respect to organizational design orientations is critical. The absence of an explicit awareness of design orientations is likely to lead to a lack of attention for human or social factors in the design of the technical system, as is shown by the partial rationality of the mechanical engineers studied by Manders (1990). In this case, technical systems are easily interpreted as being 'technocentric' or 'Tayloristic'. Of course, it remains possible that fractionized prescriptions are deliberately incorporated into technical systems, or that they make a division of labor more likely, as may be the case with the skills required by various programming methods. However, this is by no means an automaticity: the fact that a new technical system makes an extension of the division of labor possible does not necessarily mean that this option will be realized. Perhaps the main contribution of developers of human centred systems is not the development of such systems, but more the fact that they emphasize the relevance of human and social design criteria. A user-friendly programming method is by no means a guarantee for integrated jobs, but merely eliminates an obstacle for their occurrence by lowering the gap between the skill requirements for programming and operating.

In order to have technical systems fit into integrative design orientations, proponents of such orientations will need to specify the technical criteria mechanical engineers have to take into account when designing production systems. Such criteria may be general or specific for a certain design method. User friendliness is probably a technical characteristic that fits well in any integrative method. In case of MST, a specific criterion concerns the lay-out ('production organization'), which should be group-oriented. A number of small machines is preferred to one large machine, as small machines fit into group production, whereas large machines tend to be included in a functional or line organization. The technical systems have to fit in the concept of flow production (Pot et al., 1991, p. 35).

Having set out the influence of technical design orientations, a final comment on techn(olog)ical determinism can be made. In section 2.1, it was argued, that the fractionized design orientation is deeply embedded in organization design, and, as a rule, is applied, consciously or unconsciously. In that case it can be expected that fractionized jobs are likely to occur around any technical system. If the effect of the introduction of any technical system on the division of labor is studied, and the fractionized jobs have been found to be dominant, it may seem that the technical system under study, rather than the organizational design orientation, determines the division of labor. In other words: the design orientation acts as a moderator variable between the (independent) variable 'technical system' and the (dependent) variable 'division of labor'.

As argued above, technical systems as such only have a limited effect on the division of labor. The implementation of a new technical system leads to changes in the organizational task pool, which has consequences for the division of labor. This makes the introduction of technical systems an ideal moment to implement other changes. The distinction between both variables is then indeed analytical (Doorewaard and Huijgen, 1985), and it is easy to ascribe changes in the division of labor to the technical system introduced, i.e. to see some variant of techn(olog)ical determinism.

Finally, the viability of human centred systems, as well as integrative design methods, will ultimately be determined by their 'profit-centredness' (Jones, 1991, pp. 617-619). Herzog, describing the introduction of German WOP programming methods, states that earlier attempts failed right from the start because of 'the killer argument of failing economy' (1991, p. 48).

2.3 Product markets

Changes in technical and organizational arrangements are generally discussed in connection with changes on product markets. The latter are often seen as the prime motivations for implementing other changes, and hence need to be discussed here.

First, an ideal-typical picture of modern product markets is sketched. Secondly, this ideal-typical portrayal is assessed critically. Finally, some conclusions are drawn regarding these developments on product markets and their consequences.

'Modern products markets'

Many authors try to analyze contemporary developments on product markets. Several models were developed with descriptive, predictive, or prescriptive goals. The popularity of these models may be related to the suggested implications for organization, which are discussed further on. In the Netherlands, the 'phase model' of Bolwijn and Kumpe, two Philips consultants and professors of management, has become quite influential. This phase model aims to describe the subsequent phases large, multinational organizations have gone through from 1960 onwards. According to Bolwijn and Kumpe, '[l]arge multinational companies often have little freedom of choice in determining their industrial strategy' (1990, p. 44). They are forced to follow new market demands.

The model consists of four phases, which companies pass subsequently:

1 phase of efficiency, which is the traditional phase;
2 phase of quality, starting at the end of the 1960s;
3 phase of flexibility, starting in the mid-1970s;
4 phase of innovation, which is typical for the 1990s.

In every following phase, organizations have to perform according to the requirements of previous phases, plus some new requirements posed by the phase just entered. In the efficiency phase, all organizational efforts are directed at cost reduction by striving toward the mass production of standardized goods. The stable markets allowed the application of fractionized organization designs. However, the substandard product quality became problematic especially when production capacity started to exceed demand. This led to intensified competition, in which product quality became an important item. The formerly widespread belief that product quality and efficiency were opposing goals, was under attack. The efforts of quality gurus had a lasting impact, and it was realized that 'Doing things the right the first time is the best and cheapest way to run a business' (1990, p. 50); in other words: efficiency and quality are compatible, as scrap and repair costs are eliminated. Furthermore, products known for their quality gained market share at the expense of substandard quality products. In the next phase flexibility is required, i.e. the ability to produce 'wide ranges of up-to-date products' (1990, p. 46). This is only possible 'if and when efficiency and in particular quality are properly under control' (1990, p. 46). For the 1990s, Bolwijn and Kumpe foresaw the phase of innovativeness. In this phase, organizations have to be able to introduce improved products quickly. These products have to be outstanding, which is 'uniqueness' of products. The last two phases require fast feedback, flat organizations, autonomous departments, and a minimum of central staff groups. In other words: integrative organization designs are called for in order to remain competitive under the conditions described by the later phases of Bolwijn and Kumpe's model. However,

> not all companies react with the same swiftness to a new, upcoming market demand. Delay times vary. Also new market demands do not emerge for all industrial sectors or product/market combinations at the same time. So not all parts of a company will be in the same phase of our model (1990, p. 47).

The phase model proved, and still proves, to be appealing to many, even to the extent that the model's field of applicability is broadened beyond the area claimed by Bolwijn and Kumpe. The criteria efficiency, quality, and flexibility were called a 'conditio sine qua non' and 'not specific for a certain branch or a certain kind of enterprises' by Rauwenhoff, former vice-president of Philips (1986, p. 33).

The phase model is deterministic: it does not acknowledge the freedom to choose a strategy. At this point, Bolwijn and Kumpe disagree with Porter (1985), who stresses an organizations' freedom to choose a manufacturing strategy, either aimed at cost reduction or product differentiation (1990, p. 55). The deterministic character is remarkable as the model is meant explicitly for large, multinational organizations, a category one can expect (to be able) to exert considerable influence over its environment in general, and its product markets in particular. These contrasting views can be seen as manifestations of a more general debate, led in organizational theory, on voluntarism and determinism (Astley and van de Ven, 1983; cfr. section 1.1). Furthermore, the model is clearly ideal-typical. The authors acknowledge the possibility that organizations are in differing stages of development, which meets many potential objections.

The 'phase model', describing changes on products markets with subsequent necessary organizational adaptations, is by no means unique. Especially the trend from mass production toward the small batch production is noted by many authors, be it that the wording changes. Flexible specialization (Piore and Sable, 1984), the sand cone model (De Meyer and Bonheure, 1991; De Meyer and Ferdows, 1991) give similar descriptions, while the trend toward small batches is mentioned by many authors without incorporating it into a specific model (Adler, 1988; Brödner, 1987; Gunn, 1987; Kidd, 1991; Wobbe, 1991). Okko (1992) uses the word 'flexibility paradigm' in this respect.

Summarizing: according to authors mentioned above, mass production is no longer a viable product strategy for multinational companies. Instead, manufacturers are either forced, or well advised, to produce a wide and constantly changing range of high quality products. Roughly, product quality became an important issue in the beginning of the 1970s, about 1980 a wide range of products became essential as well, and a small decade later, the constant renewal of product ranges became a precondition for organizational survival.

A critical assessment

Although the widespread consensus concerning the general dominant trend in product markets can be seen as a confirmation of the 'phase model', it is hard, if not impossible, to find a methodologically sound foundation for it. Another point of criticism concerns its generalizability with respect to other periods and branches. Finally, it has to be noted that the level of analysis plays a role. Except for the first point, these criticisms concern the general trend described by these models rather than the 'phase model' itself. As will become clear in the following, these points of criticism are partially intertwined.

Bolwijn and Kumpe are explicit about the empirical basis of the phase model: it is mainly based on visits to several outstanding companies in Europe (1990, p. 44). Such a basis needs little methodological discussion. As no mention is made of the peculiarities of these factories, nor of the method of data analysis (which is clearly problematic in qualitative research), the model does not seem to have faced any scientific rigor, but is merely a systematic crystallization of personal experiences. The emphasis the model puts on 'quality', 'flexibility' and 'innovativeness', and the resulting necessary organizational changes fit well in, and at the same time have fed what seems to be the currently dominant production ideology. Its appeal might explain its uncritical use, using the model even beyond the areas it means to describe. Furthermore, given the large number of competing models, and their popularity in the practitioner's and academic world, the phase model's face value may be considerable warranting a closer look concerning the extent that the above described development from mass production to small batch production takes place. Is it restricted to large multinational enterprises as Bolwijn and Kumpe state, or is it of use for other firms as well?

Emery and Trist's (1965) classic article 'The Causal Texture of Organizational Environments' serves as the starting point for the remainder of the discussion. In this paper, originally read at a conference in 1963, Emery and Trist identify four ideal types of 'causal texture', i.e. the influence of the environment in general on organization structure. This conceptualization of the environmental role was motivated by a number of cases, one of which is described at the beginning of the article. The case concerns a British vegetable canning company, that had held a stable 65 per cent market share since before the war, and had decided to invest in a new, rigidly automated factory set up for mass production. When the factory was being built, a number of small canning firms started to market cheaper brands, initially as a means to reduce capacity underutilization due to the seasonality of their imported fruits businesses. These small firms could cheaply procure American surplus crops, which could not be processed by American producers of quick-frozen foods. British consumers now had a greater variety to choose: more canned products, quick-frozen food, and an increased number of substitutable vegetables. Furthermore, the upcoming grocery chains started to market cheap house brands, for which bulk orders were placed with the small canners. Unfortunately, the new factory could not be adapted, and as a result of all these intertwined developments the company's market share dropped significantly. If nothing else, the canning case shows that this firms' mass production strategy was no longer viable as early about 1960 (the data must have been gathered in the beginning of the 1960s).

Emery and Trist try to grasp these environmental influences in the notion of the 'causal texture of the environment' in four ideal types. The first three types

> have already, and indeed repeatedly, been described - in a large variety of terms and with the emphasis on an equally bewildering variety of special aspects - in the literature of a number of disciplines, ranging from biology to economics and including military theory as well as psychology and sociology. The fourth type, however, is new, at least to us (1965, 24).

The four types are:

1. placid, randomized environment;
2. placid, clustered environment;
3. disturbed-reactive environment;
4. turbulent fields.

Although it is clear from their article, that the intensity of environmental changes necessitates a similar intensity of organizational responses, consequences for organizational structures are only superficially discussed. What they do emphasize, however, is that 'turbulent fields demand some overall form of organization that is essentially different from the hierarchically structured forms to which we are accustomed' (1965, pp. 28-29).

Later, the four types were supplemented by a fifth, called the 'hyperturbulent' environment' (McCann and Selsky, 1984) or the 'vortical' environment (Babüroglu, 1988).

The Emery and Trist paper can be interpreted as an early indication that mass production was no longer a viable market strategy by the beginning of the 1950s, and that changes of differing intensities were occurring. A more elaborate discussion of the organizational consequences of such changes can be found even earlier in the work of Burns and Stalker (1961). Based on work of previous authors and their own case-studies in British electrotechnical firms, Burns and Stalker distinguished two polar types of organizations, namely 'mechanistic' and 'organic' systems of management, and, as a new element, related these systems of management to environmental characteristics. Given a particular environment, both types are rational. The mechanistic management system is appropriate to stable conditions, and is characterized among others by a detailed division of labor, tight hierarchical structure and matching specifications of each individual's rights and obligations. This bureaucratic way of organizing is not found in the organic structure, which is appropriate in changing conditions. These give constantly rise to unforeseen problems, which are hard to resolve by the mechanistic way of working. Characteristics include a network structure of control and authority, and continual adjustment and redefinition of individual tasks.

With respect to the timing of the developments, the experience of Olivetti, one of Philips' competing large, multinational enterprises, is illustrative.

> The character of Olivetti's market has changed in fact; the market for microcomputers, data processing systems, numerically controlled machine tools, and so on is more specialized, more differentiated in terms of performance requirements, more demanding in terms of technical quality, and, above all, subject to stronger competition. To be successful in these markets the company must not only offer competitive prices and very high quality; chiefly, it must also supply a wide range of products and constantly improve them to meet customers' requirements. As a consequence, for example, the company's catalog grows larger, since each product is supplied in an increasing number of models, and the products' lifetimes are considerably shorter (Butera, 1975, pp. 167-168).

Interesting common features of these two examples are that both concern electrotechnical firms, and both describe the development away from mass production. Especially the work of Burns and Stalker suggests that as early as the late 1950s, electrotechnical firms were operating with different organization structures in different product markets. Some 15 years later, Olivetti seems to have felt a need to adapt its organization to changed product markets. Naturally, both examples are illustrations and do not provide any proof on the timing of the suggested general trend toward smaller batches. Using these examples as a proof would provide a wobbly empirical base, as no systematic, large scale investigation concerning market developments has taken place. The same, however, holds for the suggested trends, which are often based on cases and may not pretend generalizability (Yin, 1989), or miss a systematic account of data collection, analysis and interpretation, which makes the findings subject to vulnerability (Lammers, 1986, p. 19).

Another issue concerns the branches in which these market developments take place. The electronical and automobile industries are two examples. Yet, branches with high proportions of small and medium-sized firms such as the metal products and machine-building branches are typically not covered by the phase model. As far as strategies with a high product variety, a high product variability, and small batches, are the traditionally most common strategy in particular branches, as is the case for the Dutch metallurgic industries (Stokman, de Groot and Leurs, 1989, p. 20), branches such as the electronics and automotive industries are merely moving in the direction which is already familiar in other branches (Manske, 1987, p. 132). Within the chemical industries, a distinction is made between bulk chemicals and specialties, thereby allowing for different markets strategies.

To be brief, inter- and intra-branch differences in product strategies occur. These may be related to organizational size, although this relationship is also not universally assumed, as can be shown by quoting Sorge and Streeck: 'Small batches may be beautiful in small or large units, with small or large volumes alike. The same applies if beautiful is replaced by ugly' (1988, p. 29).

A third comment concerns the level at which these developments take place. At the corporate level, an increase of product variety can be achieved by acquiring another company, but this will probably have no impact on the organization of work at the shopfloor. The global organizational product and market strategy of (in Bolwijn and Kumpe's case large, multinational) organizations on the one hand, and the output characteristics at the shopfloor level have to be decoupled to investigate the effect at the shopfloor. Although both levels, perhaps connected by several additional layers, are not unrelated, they have to be analytically separated. A greater product variety may occur at both levels, but the level of analysis is important to determine the organizational consequences.

Bolwijn and Kumpe, in a publication co-authored with three other Philips employees, assert that products tend to consist of fewer components. The reduction of product complexity, as well as modular product design[5] are even prescribed in the same volume (Bolwijn et al., 1986). In case of modular design, the combination of a limited number of standard components can result in customized designs. In other words: the characteristics of end products may be totally different from those at the component level. If a strategy of supplying a broad product variety is combined with modular product design, shopfloor conditions may remain unchanged, whereas the distribution and sales departments will have to cope with an increased product range. Another example is group technology, in which similar parts are identified and grouped together into part families. As the individual parts are largely similar, they have to be processed in similar ways too, which makes processing efficiencies possible (Groover, 1987, pp. 433-456). According to Tidd and Francis (1992, p. 808), American and many European manufacturers, in contrast to their Japanese counterparts, have consistently reduced product ranges to facilitate automation and to realize cost savings.

Summarizing the point so far: the problem is not so much that the 'modern' models do not capture the full reality of developments on product markets, but rather that strategies that can be considered to be alternatives to mass production such as focusing on product quality, market niches, or broad product ranges, existed prior to their baptism. This points to a less static situation some twenty-five years ago than is often assumed. After describing the dominant concept of mass production, Skinner argues that '[t]oday these factors are changing, with the extent and rate of change varying with the industry and the company [...] The assignment becomes - "Make an increasing variety of products, on shorter lead times with smaller runs, but with flawless quality"' (1966, p. 140). If this quotation of Skinner adequately describes the situation in the USA in the mid-1960s, how could it take twenty years before it had an effect outside the USA?

The unconscious or conscious neglect of older sources, together with personal interests, easily leads to tendentious opinions concerning product strategies, often vigorously propagated by their 'inventors', and uncritically adopted by their disciples. Nevertheless, the multitude and popularity of different models indicates that something is happening in the area of product strategies, which has consequences for the efficiency with which organizations operate. In the following, an attempt will be made to identify the main elements of a product strategy as far as these influence the division of labor around stand-alone machines, as these are the focus of this study.

Output characteristics and organizational characteristics

Sorge, in what seems to be an effort to revitalize the contingency approach, distinguishes the following output characteristics:

- Volume of output
- Product scope, which is divided in product range and product variability
- Product complexity (1989, pp. 24-27).

Product scope expresses the overall product variety, while its dimensions product range and product variability refer to respectively the product variety handled by organizational sub-units, and the product variety handled within a particular sub-unit. In these terms, the product scope can be increased as a result of diversification, i.e. an increasing product range, or by a larger number of product produced by a particular unit, i.e. an increasing product variability. Product volume is, as product range, defined at the organizational level.

If the focus is on departments or the shopfloor, these output characteristics have to be modified. Clearly, product variability in the above definition is only meaningful above the departmental or unit level. Another necessary modification concerns a dynamic aspect, namely the extent to which changes occur in these output characteristics. For this, the now obliterate term 'product variability' seems most appropriate. Furthermore, a remarkable absentee is 'batch size', which is placed under the cover of output volume. In Sorge and Streeck (1988, pp. 25-29), batch size is attributed an important role as apparently the main link between properties of product markets and organizational characteristics. Further on in this article, it is treated as a synonym to output volume. Whereas such an equalization of output volume and batch size may be permissible at the organization level, at the shopfloor level output volume has to be unravelled as batch size and repeat frequency. Summarizing, at the shopfloor level the following output characteristics have to be taken into account:

- Product complexity.
- Product variability.
- Product range.
- Batch size.
- Repeat frequency.

By using Sorge's concepts 'workflow differentiation', 'workflow continuity' and 'workflow rigidity', the relationships between these characteristics and the organization of labor can be described. Workflow differentiation refers to the extent to which the workflow is differentiated into discrete jobs, machines or plant components. These can be conducted simultaneously, which is 'parallel differentiation', or subsequently, 'sequential differentiation'. The higher (end) product complexity, the larger the number of operations, and the higher the number of possible combinations of operations, the higher the possible variety in workflow differentiation. Working stations, whether mechanic or human, can be grouped sequentially in a production line or group, or parallel in a functional lay-out. In the latter case, machines performing similar operations are grouped, whereas in the other two cases all necessary work stations to process a (group of) product(-s) are placed together. Within the same production department, different lay-outs may occur simultaneously.

'Workflow continuity' refers to the degree that the workflow is interrupted by the total time required for converting the production system to other products, although Sorge acknowledges that interruptions can also be caused by breakdowns, maintenance, and organizational slack. In cases of a high repeat frequency or a high product variability, the workflow is frequently interrupted. The length of the conversion time, or 'set-up time', is generally in the first place a technical characteristic, although organizational measures may also lead to significant reductions (Schonberger, 1982).

'Workflow rigidity' denotes the extent to which a production unit is restricted to operations for particular inputs, transformation processes or outputs. The higher the product range and variability, the higher the need for production factors that can be deployed for a variety of operations. Workflow rigidity can be observed at several levels of aggregation, starting from individual employees c.q. stand-alone machines, and their combinations in the form of organizational structures. Conventionally, a high utilization grade of production factors is stressed which is visible in the common accounting practices. In this case, an efficient utilization of production factors is stressed. The higher the product range, variability, and repeat frequency, and the smaller batches, the greater the technical difficulties to design and manufacture machines suited to handle such a product diversity, and hence their price. Dedicated machinery typically handles a small product range, which is expected to remain stable for a substantial period (the canning case being a typical description of the results of this assumption being invalid), with large batches and a low repeat frequency. In this case, workflow continuity can be realized by using a line lay-out, in which all working stations are closely coupled sequentially. Table 2.4 contains some ideal-typical combinations of output characteristics and lay-out.

Table 2.4
Output characteristics and lay-out

Lay-out	Product variability	Product range	Batch size	Repeat frequency
Line	very low	small	(very) large	low
Group	low - intermediate	small	intermediate - large	high
Functional	intermediate - high	broad	varying	varying

A line lay-out is only (economically) feasible in case of a special combination of output characteristics. Total production volume has to be sufficient to earn back the (normally substantial) investments in dedicated equipment. In case of a group lay-out, and even more in case of a functional lay-out, more variety in the output characteristics is possible. By means of a group lay-out, related products, 'product families', can be handled, which makes this lay-out less dedicated than a line, but more dedicated than a functional lay-out. Compared with the latter, the lower variation in output characteristics leads to a lower need for coordination. Furthermore, the situation is more easily controllable as product flows are less complicated in a group lay-out than in a functional lay-out. Typically, the complex product flows in a functional lay-out, combined with the emphasis on high degrees of machine utilization, make such production systems intransparent and hard to control, which is an important reason to plea for the grouping of related products into product groups in modern sociotechnology. However, total production volume has to be sufficiently large in order to prevent low utilization degrees.

In organization and industrial sociology widespread support is found for a coupling between output characteristics and organizational structure, which is a classical contingency argument (Sorge, 1991, p. 165). Burns and Stalker's mechanistic and organic systems of management typically fit stable respectively dynamic output characteristics. In the terms of section 2.1 integrative designs, which structures require less coordination than fractionized designs, are most efficient if there is a high need for coordination, thus in case of dynamic output characteristics. In case of the specific characteristics mentioned in Table 2.4 the fractionized line organization is most efficient as economies of scale can be achieved.

Following the reasoning above, however, this coupling may be less tight than generally assumed. Modular product design may link a low product range and variability, and large batches in component production, in other words ideal conditions for a fractionized design, to unique, and even customized end products. Furthermore, the grouping of related products into product families, as prescribed in MST and group technology show that it is possible to create different organizational arrangements starting from similar output characteristics. Finally, in MST the lay-outs mentioned in Table 2.4 are typically combined with specific design orientations. The group lay-out, which is to be combined with task groups, is characterized by an integrative design, while both the line and the functional lay-outs are typically linked with fractionized designs. Whereas such a combination seems logical for the line production, it is less evident to see why the functional lay-out, typically handling a large variety of products, i.e. operating in a dynamic environment, ought to be characterized by a fractionized design. One possible explanation can be found in a supposed high pressure on machine utilization in a functional lay-out, which leads to fixed man-machine assignments, but such an explanation seems somewhat artificial. Another explanation might be, that the functional and line lay-out are both traditional ways of grouping machines, which are likely to be combined with the traditional fractionized design orientation, but again this argument is hardly convincing. Only the line lay-out fits well in the contingency argument. If Table 2.4 is correct in linking output characteristics and lay-outs, one may expect, following the contingency argument, that both design orientations can be found in functional and group lay-outs, depending on the output characteristics in specific cases. Such an argument also allows for an internal division of labor within task groups, characterized by either an integrative or a fractionized design orientation (Eichener, 1991).

Conclusions and discussion

The output characteristics 'product complexity', 'product variability', 'product range', 'batch size', and 'repeat frequency' are related to characteristics of organizational structures. Depending on the scores on these variables, and the specific combinations of these variables, some lay-outs and design orientations are more likely to occur than others. This is not to say that output characteristics necessarily have to be the starting point of the discussion. It is just as well possible that organizations possessing specific characteristics adapt their product market strategy in order to fit the strategy with the structure, rather than the other way around. Put this way, whether 'strategy follows structure' or 'structure follows strategy' is less relevant than that strategy and structure fit, which can be described as an 'elective affinity' (Sorge and Streeck, 1988). Another element that is beyond the area of interest here concerns the question whether organizations have a certain freedom of choice with respect to their market strategies as Porter argues, or that they are forced to follow their competitors as Bolwijn and Kumpe are inclined to believe. Perhaps both are possible simultaneously (Hrebiniak and Joyce, 1985). However, given the large variety of options presented above, the more deterministic options seem unlikely.

For a discussion of the influence of output characteristics at the shopfloor level, it is essential to identify the relevant output characteristics at this level. Most of the models portraying the ideal-typical market developments concern the organization level, which, as argued above, may be decoupled from the shopfloor in various ways. These include the distinction between organizational sub-units, modular product design and the grouping of products. Even if there is a close coupling, it remains to be seen whether, or perhaps more accurately, to what extent the market developments follow the models designed to describe them. The large diversity of these models leads one to believe that mass production is not the dominant product strategy, but whether this is a breach with the tradition, or the tradition itself, is unclear due to lack of longitudinal statistical data covering only a selection of industrial branches. The impressionistic evidence in favor of these models can be supplemented by equally impressionistic evidence demonstrating the opposite. If changes are actually occurring, they are likely to have an effect on organizational structures, but this effect can be expected to be quite moderate.

2.4 Labor markets

This section deals with the relationship between skill level needed to fulfil jobs offered by employers on the one hand, and the skill level possessed by of would-be employees on the (external) labor market on the other hand. The issue has a qualitative and a quantitative dimension. The qualitative dimension concerns the degree to which both skill levels fit. The quantitative dimension concerns the question whether or not the demand for labor equals the supply of labor, which influences the qualitative dimension. To avoid confusion, demand of labor refers to the jobs offered by employers to employees, and supply of labor refers to the labor capacity offered to employers by employees. 'Skill level' refers to the demand side, i.e. the skill level needed to fulfil a particular function; 'education level' refers to the supply side, i.e. the skills employees actually have acquired.

In many industrialized countries, compulsory mass education has led to a rising general level of education, which translates itself in a rising education level. This phenomenon has been mentioned as an important reason to introduce integrative design methods. For instance, MST-foreman de Sitter states: 'The qualitative tensions on the labor market, which are shown in the differences between the supply and demand, will have to be bridged by changes in the production' (1981, p. 112). Similar prescriptive arguments, although not always directed at organization design, can be found in literature on 'human capital' and to use more recent terminology 'human resource management' (HRM). In the foreword of the book describing the 'groundbreaking' approach of the Harvard Business School to HRM, Beer et al. state that 'The evidence seems to indicate [...] that in many corporations, human resources are underutilized and underemployed [...] A more highly educated work force has challenged management to find ways of increasing employee involvement, responsibility, and participation in the enterprise' (1984, p. viii). Human resources are seen as social capital which has to be developed and maintained. The economic logic underlying these claims is that employers have to make full use of the capabilities employees possess, i.e. their education level. The argument does not stop with the current, static situation, but often also includes a dynamic perspective stressing the development of employees' potential skills by means of (continuous) schooling. In an economic perspective, the gap between the education level and the skill level is seen as an untapped source of profits. In humanitarian and psychological approaches, this gap is interpreted as insufficient (opportunities for) self-realization of individuals. Both are undesirable, and thus employers as well as employees might benefit from a full utilization of the individual's capacities. Parallel to the development of integrative design methods in section 2.1, a convergence of employers' and employees' interests may emerge. This parallel can be seen as an implication of the application of integrative design methods for the organizational function structure. By applying the principle of a minimal division of labor, functions get composed of a greater variety of tasks, requiring more and often higher skills.

The economic logic underlying the contrasting fractionized model of utilizing labor was already spelled out in section 2.1 by referring to the Babbage-principle. Different tasks require different skills, and are often rewarded differently. This last practice is institutionalized in the use of function classification and renumeration systems. In the Netherlands these are increasingly being used, with more than fifty per cent of all employees in organizations employing more than 19 persons being classified in 1989 (Loontechnische Dienst, 1991). Under influence of this argument and the ones listed in section 2.1, the fractionized organization gained ground, with the consequence that many jobs were created that require few or skills. This existence of 'everybody's tasks' (*Jedermannstätigkeiten* in German) was repeatedly mentioned (van der Waerden, 1911; Kuylaars, 1951; de Sitter, 1981).

The application of design principles leads to a certain organizational function structure, which can be defined as 'the total of all functions existing within an organization' and forms the demand of an individual organization for specified categories of labor. The macro-economic 'qualitative employment structure' can be found by summarizing the demands of all individual organizations.

In section 2.1, labor market shortages were mentioned as an important reason to experiment with work structuring, which exhibited integrative design orientation. It has to be observed, however, that these labor market shortages were economy-wide, and were felt to **necessitate** integrative experiments. In the absence of such shortages, other mechanisms are likely to work. Gerlach very explicitly states the economic advantages employers expect from hiring overqualified employees:

> An expected greater labor discipline and quality (compared to uneducated employees), shorter periods to learn the job, a (suspected) close availability of practized skills and qualifications, acquired during the education, as well as a potential broader deployability in the sense of

flexibility (1990, p. 167).

Gerlach's project was carried out in three German areas, with different degrees of unemployment. In case of unemployment, skilled personnel was forced to accept functions under their education level. This phenomenon was related to the level of unemployment: the higher the unemployment, the higher per centage of employees working under their skill level. When the unemployment rate went down some years later, much of the skill level had become obsolete. Gerlach concludes, that the shortage of skilled employees in Germany was partially caused by these personnel policies of the employers.

Obviously, this strategy of hiring over-qualified employees is in sharp contrast with the strategy over avoiding under-utilization.

Unlike the application of design methods and use of output characteristics, statistical data concerning the developments of skill and education levels in the Netherlands are available, which enables to draw conclusions with a sounder empirical basis and less speculative nature than those in the previous sections.

Empirical data about the Dutch labor market

In a summarizing article, Huijgen (1990) provides some data concerning this trend in the Netherlands. In Tables 2.5 and 2.6, 'skill level' was measured on a scale from 1 to 7, with level 1 representing the lowest level (unskilled jobs). Skill level 2 represents semi-skilled jobs, and the remaining levels contain different levels of skilled jobs. The highest two levels are not represented in Table 2.6, as no blue collar manufacturing jobs exist at these levels. From Table 2.5 it follows that the average economy-wide skill level has risen over the period 1960-1985. In other words: 'regradation' or 'upgrading' has taken place. However, controlling for the nature of the jobs one finds, that blue collar jobs were degraded in the same period. The same conclusion can be drawn for white collar jobs up to 1977; after 1977, the average skill level of white collar workers rose to end at approximately the level of 1960. The regradation of total employment is caused by the fact that white collar jobs, which on average and consistently score almost two points higher than blue collar jobs, came to make up an increasing proportion of total employment (last columns of Table 2.5).

Table 2.5
The Dutch qualitative employment structure from 1960 to 1985

Year	Average skill level				Per cent of employment	
	1960	1971	1977[6]	1985	1960	1985
Blue collar jobs	2.71	2.58	2.54	2.53	61	39
White collar jobs	4.41	4.37	4.23	4.39	39	61
Total employment	3.36	3.47	3.47	3.67	100	100

Source: Huijgen, 1990, pp 100-101

In Table 2.6, a closer look is taken at blue collar jobs in manufacturing between 1977 and 1985.

Table 2.6
Qualitative employment structure for blue collar jobs in Dutch manufacturing industries, 1977-1985 (absolute numbers)

Skill level	1977	1985	Index
1	143500	143500	100
2	245400	162300	66
3	184400	143300	78
4	154900	133900	86
5	5900	11400	192
Total	733900	594400	86

Source: Calculated from Huijgen, 1989, p 85

Two findings from the Table 2.6 need to be discussed. In the first place, the average skill level of blue collar manufacturing jobs rose from 2.46 to 2.52 between 1977 and 1985, which points to regradation. However, a closer analysis of Table 2.6 reveals that this rise is caused by an increasing number of jobs at the highest level. In relative terms this increase is quite spectacular, but in absolute terms this level still makes up a moderate proportion of all blue collar manufacturing jobs (about 2 per cent). The number of jobs at the remaining levels all shrank **with the exception of the lowest level**. In relative terms, polarisation has taken place between 1977 and 1985. Using wage levels, Kloosterman and Elfring (1991, pp. 148-149) find a similar trend for the Dutch manufacturing industries in general for the period 1979-1988.

In the second place, total employment in this category fell from 733,900 jobs in 1977 to 594,400 jobs in 1985, while economy-wide total employment rose from 4,087,000 to 4,374,100 jobs in the same period. The falling employment levels for blue collar jobs in manufacturing can be explained by several factors, job displacement through the use of new technical systems (mechanization and automation), an increasing labor productivity, and an economic recession being the three main reasons.

Whereas these two findings are unambiguously demonstrated by statistical data, their explanations are not evident. Although ongoing mechanization has undoubtedly taken place, one would assume that this would lead to a loss of unskilled jobs (skill level 1), as simple, often direct tasks are most easy to automate. Yet the number of level 1 jobs has remained constant. Thus, the effect of job displacement may have been off-set by the creation of unskilled tasks and jobs. If more complex tasks were created due to mechanization, and combined into functions at the intermediate levels, this tendency would have compensated for job losses at those levels. The latter may have been caused by bankruptcies, possibly in relation with the ailing Dutch economy, in branches with high proportions of people working at these levels. Yet the causes of the developments in the qualitative employment structure need further research. One conclusion that can be drawn is that the development and application of integrative design methods such as work structuring and modern sociotechnology, both of which seem to have focused on manufacturing, has not led to a general rise of the skill level in manufacturing. Unlike unskilled jobs, semi-skilled and, to a smaller degree, skilled jobs have been eliminated in manufacturing. If nothing else, the data of Table 2.6 do not provide any indication that the fractionized design orientation has lost its popularity, nor that the integrative design orientation started to break through, as Kern and Schumann (1984) concluded for some German manufacturing industries in approximately the same period. In this last respect, it is striking to find that six out of the total of nineteen manufacturing industries showed a larger increase of the skill level from 1977 to 1985 than the economy-wide average (which is far higher than the average for manufacturing industries). These six include all three sectors included in the study of Kern and Schumann (1984), which leads one to wonder whether these three branches represent exceptions rather than the rule.

Concerning the supply side, it has to be noted that the Netherlands had comparative high unemployment rates in the 1980s. Next to unfavorable business conditions, this may have been a barrier for the application of integrative designs. In general, the availability of personnel no longer necessitated integrative experiments or other methods to secure the workforce, and 'crowding out' started to occur. Crowding out refers to the fact that people with higher education levels than needed for specific jobs accept these jobs, with the result that persons with fitting education levels have to accept lower qualified jobs or stay unemployed. Crowding out is likely to occur when the supply of labor is greater than the demand for labor at the higher segments of the labor market. The confrontation of skill and education levels, leads Huijgen to conclude that crowding-out has taken place (1989, p. 46). Huijgen's 1989 study generated some disputes in Dutch literature, mainly concerning the interpretation of his empirical work (Huijgen and Smits, 1990; Salverda, 1990; Teulings, 1990; Wielers and Glebbeek, 1990).

According to Salverda (1990), the data can not solely be interpreted as underutilization, but also as decreasing overutilization. Especially before 1945, many employees had not been able to complete a study fitting with their capabilities, but proved to be able to fulfil jobs with a higher skill level than these persons' education level. A second reason for overutilization is a lack of adequately trained personnel. Huijgen and Smits (1990) responded to his comment that the empirical data do not allow to draw any conclusions as they do not contain data on (formally or informally) acquired skills after the completion of the formal education. Another point made by Salverda is that crowding out must lead to unemployment at lower education levels. Huijgen and Smits (1990) reject this, because employees at these lower levels may displace people below their level. Given full employment within the economy as a total, crowding out and underutilization may still occur.

Wielers and Glebbeek (1990) bring several points into the discussion, all pointing to advantages of a rising education level. In the first place, the education level, as measured by the highest level of formal schooling reached, is an indication of the individual's level of knowledge as well as an indication of this person's ability to learn. Employers often (have to) provide additional schooling to employees. By recruiting employees with high education levels, employers can economize on schooling costs as these employees are expected to learn quickly. Secondly, drawing on the hedonic function theory, a function can be performed better (for instance measured by the individual's labor productivity) by employees with a higher education level. In the third place, the introduction of innovations is facilitated by the rising education level.

Teulings (1990) makes a distinction between crowding out and substitution. The first phenomenon is caused by differential skill levels, while the latter is caused by differential wages. As skill levels and wages are generally directly or indirectly related, and the Dutch wage structure is fairly rigid, this distinction is in the first place of theoretical interest. Both have the same consequence with respect to underutilization.

Despite their differences of opinion, all authors agree on the fact, that crowding out has taken place on a large scale in the Netherlands during the first half of the 1980s, and that this phenomenon is made possible by a shortage of jobs at the higher segments of the labor market. These empirical studies show that the continuous pleas to utilize human capital have only sorted a limited effect.

The interpretation of underutilization clearly differs. In a human resource perspective it is clearly an undesirable phenomenon. Wielers and Glebbeek, on the other hand, give several reasons why such a qualitative tension may have positive economic spin-offs in the form of lower schooling costs, higher job performance, and higher innovative potential. To this list Gerlach (1990) adds that overqualified personnel may form a pool which facilitates the (possible) future implementation of 'New Production Concepts'. On the other hand, underutilization may have negative effects on employee behavior and attitudes, as discussed in section 2.1. Taylor even prescribed underutilization in case of varying work: 'if the work to be done is of great variety [...] the men engaged should be too good for their jobs' (1903, p. 142). However, for two reasons integrative designs may also lead to a certain underutilization.

In the first place, broad employee deployability goes hand in hand with functions consisting of a broad spectrum of tasks. The greater the deployability, the higher the chance that the skills needed to perform these tasks diverge. In the fulfilment of their functions, employees only have to make a limited use of skills needed for the most difficult tasks, i.e. there is a danger of creating underutilization combined with a broad deployability. The redundancy of functions ('overcapacity of functions in each person within the organization'; van Eijnatten, 1991, p. 5), prescribed in sociotechnical systems design, contains this potential problem (also see Morgan, 1986, pp. 98-99). Furthermore, if the existing function renumeration systems value functions based on the highest skill level needed for any task in the function, the achievement of a broad deployability leads to a negative direct economic effect as the function is valued higher than would have been necessary if the simpler tasks were combined into separate functions, i.e. if the Babbage-principle were applied.

Secondly, integrative designs lead to rising skill levels, but recruiters may overestimate this impact. Steijn and de Witte (1985, pp. 1106-1110) demonstrated that personnel managers responsible for recruiting **perceived** technical and organizational changes to necessitate higher skill levels. The tendency to signal rising skill levels was confirmed in a study by the branch organization in the metallurgic industries FME. On the general question whether education requirements had risen or fallen in the last few years, 45.4 per cent responded 'higher', 51.1 per cent 'equal', and 2.7 per cent 'lower' (FME, 1990c, p. 15). As Steijn and de Witte notice, these higher requirements are not necessarily connected with a higher average job content, although this was perceived to rise as well. The personnel managers denied that the rising requirements were caused by the situation of high unemployment, but were due to upgrading. Yet such a conclusion is hard to combine with the statistical data presented in Tables 2.5 and 2.6. The rise in skill requirements may be due to perceived rather than actual upgrading. The perceived effect may be larger if integrative designs are actually implemented, but is even present in the absence of changing job contents.

Personnel shortages and design orientations

Whereas the prescriptive plea to increase the qualitative dimension of the demand for labor contains a voluntary element in case of an abundant supply of labor, this situation changes in case of labor market shortages. In a tight labor market employees can be expected to be able to acquire jobs that fit with their skills. Underutilization is therefore not likely to occur in tight labor markets. This argument has to be seen in the light of segmented labor markets. Whereas a general scarcity of employees might result in fitting jobs for employees of all skill levels, the pattern changes in case of scarcity at some segments of the labor market. Given the fact that employees are able to perform tasks of a lower skill level, but are not likely to perform tasks above their skill level, scarcity is most likely to occur at the higher segments of the labor market.

In his evaluation of integrative experiments in the Netherlands, Ramondt (1974, p. 114) discusses the role of personnel shortages. Work structuring is seen as an instrument to improve an organization's position on the labor market, both to recruit and to retain scarce personnel. Work structuring was aimed at labor requiring low skills such as assembly line jobs, thus at the market for unskilled work. However, Ramondt is sceptical about its effectiveness: 'Lowly qualified work is interchangeable to a great extent. Not the content of jobs, but wages are decisive for the [employee's; JB] choice for an employer. It is hardly possible to add aspects to work content, which can withhold employees from their market orientation' (1974, p. 114). Higher wages may compensate for the unattractive aspects of this work.

Perhaps Ramondt's scepticism is somewhat exaggerated. Although offering higher wages is definitely a less strenuous way to improve ones labor market position than restructuring jobs, some arguments can be raised in favor of the latter policy. In the first place, higher wages increase labor costs, which is exactly the opposite of an important goal of fractionized designs. Secondly, legal constraints, function classification and renumeration systems, and fears to start a 'wage war' may limit the possibilities to increase wages. In the third place, improving an organization's labor market position has repeatedly been mentioned as an important motivation for integrative experiments in Sweden, a country with a persistent low unemployment. Fourthly, both measures may be taken simultaneously, which is even likely in case of coupling required skills and pay.

On the other hand, increasing unemployment may alleviate or even resolve recruitment problems more or less automatically. Huijgen (1984) mentions three types of problems, that Dutch manufacturing enterprises with simple production encountered in the 1970s, namely:

1 recruitment problems;
2 problems with respect to employee behavior, including resignations, absenteeism and sabotage;
3 organizational dysfunctions, due to employee behavior and organizational complexity;

With increasing unemployment the recruitment problems are alleviated, but the other problems remain.

Work structuring and other integrative methods have been used to make jobs more attractive, which is especially relevant in those cases where the problems of fractionized organizations are most pressing, i.e. with jobs at the lower levels in the organization. Whereas a tight market for unskilled work stimulates the use of integrative methods at the lower end of the labor market, at the higher segments, personnel scarcity can have another result. This scarcity at a medium segment of the (technical) labor market may result in a higher rather than a lower division of labor. The market for technicians with vocational training may provide as an illustration. One scarce category are CNC-programmers. The employers' organization for the metallurgic industries FME states in a report about the labor market: 'MTS'ers resign if they have to perform pure production work. ('they want to program and be prepared for functions in middle management')' (1990, p. 15). Employers may see themselves forced to offer programming tasks to MTSers, if they want to keep the MTSers. The more mundane tasks will then remain for lower qualified employees. Another example illustration is found in Dutch health care. Out of a range of possible solutions, 'vertical differentiation of functions', splitting off simpler tasks from functions and combining them into newly created, low level functions, is sometimes used (Vissers et al., 1991, pp. 40-44), sometimes even recommended to alleviate recruitment problems.

Conclusion and discussion

Several conclusions can be drawn with respect to the relationships between labor organizations and the labor market. Firstly, the argument that changes in the labor organization are necessary because of the rising general level of education has to be considered as normative and prescriptive, rather than positive and descriptive. During the 1980s, the gap between education levels and skill levels has widened. This phenomenon is not necessarily interpreted as negative: it has been argued that overskilling leads to an increasing productivity. The increase of this gap was possible because of high unemployment levels, allowing employers to be very selective in the choice of their employees. This situation may change of course as labor markets tighten, as developments on the qualitative and quantitative dimensions of the labor market are interconnected. In that case, employees find themselves in the comfortable position of being able to choose the employers according to their preferences.

Secondly, one of the elements playing a role in this choice is the job content offered, certainly if this is coupled to the wages via function classification and renumeration systems. A redesign of labor organizations is only likely to occur in case of a general scarcity of employees. In case of scarcity at some labor markets segments, the employees in these segments are likely to acquire jobs that fit with their skills at the expense of lower qualified employees, for whom the more mundane tasks are left. Hence, labor market shortages may prevent rather than stimulate the diffusion of integrative organizations.

Thirdly, a polarization in the qualitative employment structure (at the macro-level) of blue collar manufacturing jobs has taken place in the Netherlands between 1977 and 1985. This is caused by the net elimination of jobs at the intermediate levels, with a constant or increasing employment at the extreme levels. This phenomenon is remarkable, given that one would expect to see the impact of an integrative design method such as MST precisely at the lowest level.

Finally, the perceptions concerning skill levels of personnel managers in their role as recruiters are essential for the actual skills asked. Technical and organizational changes seem to nurture the perception that skill levels are rising, which leads to rising requirements posed on the education level of applicants. The application of integrative design principles may strengthen this relationship, possible even widening the gap between skill and education levels.

A prospective lookout concerning future labor market developments is, by definition, speculative. Demographic developments, the European common market, the worldwide economic development, and the continuously increasing labor productivity are only four out of an almost endless list of factors affecting the demand and supply side of the labor market. The *communis opinio* for the short and medium term concerning the Netherlands, however, is that personnel shortages are likely to continue to exist, or even increase, at certain segments of the labor market. Personnel with lower and especially medium level vocational training are among these segments. Perhaps, this will result in a further polarization of the employment structure of manual workers in Dutch manufacturing.

2.5 Research model and hypotheses

In 1972, Child published an article in which he criticizes structural contingency theorists for neglecting decision making processes. Child sees the organizational structure as the resultant of decisions taken by the power holders in the organization, the 'dominant coalition'. He argues that the structural approach in which patterns of association between environmental and organizational characteristics are sought for, ignores the political processes, in which decisions concerning the future courses of action are taken. These 'strategic choices' are made by the 'dominant coalition', i.e. those organizational members who hold the power to make this kind of decisions (generally construed as the 'management'). The actions to be taken may concern the internal functioning of the organization, but also the manipulation of environmental factors or the selection of the environment to operate in. Child's analysis allows to take the interests and strategies of different parties within the organization into consideration, which is a pluralist point of view. Child's position is also voluntarist in that the parties involved are relatively free to take actions they perceive to be appropriate. The notion of design orientations introduced in section 2.1 fits into such an analysis, as they may influence the perceptions of the dominant coalition about the structure of the labor organization. Environmental conditions may put constraints on the long-term effectiveness of these strategic choices in the sense of organizational continuity, but in principle the dominant coalition is free to choose. This voluntarist standpoint allows a loose coupling between organizational structure and environmental conditions, as consequences of developments in the organizational environment (for instance, input and output markets, political situation, legislation, and economic prospects) have to be identified and later translated into appropriate actions (often via an internal political process). If a certain environmental development is overlooked, one may find a relationship between this factor and the organizational structure which does not seem to make sense.

Child's focus on the process of decision making may not be incompatible with the position of the structural contingency theorists but can be seen as the other side of the same medal. Child focuses on decision making processes, where the structuralists studied the results of these decisions. When a large number of organizations is studied and all these organizations are scored on certain variables, a pattern between independent and dependent variables is likely to be found, irrespective of issues such as the underlying political processes leading to the formation of this pattern and whether or not this pattern could have been different, i.e. whether or not a certain degree of freedom exists. In a decision making perspective it can be interpreted as the dominant choice made by the majority of dominant coalitions within the organizations involved in the study. These decisions are postulated to be intentionally rational, i.e. in the interest of the dominant coalition according to the best judgments of this coalition. Such a rationality is not necessarily aimed at profit maximization as assumed in economic theories, but the organizational performance has to meet minimum profitability requirements in order to guarantee its long-term continuity. The intentional rationality also applies to the design orientations adopted.

Another issue that needs to be discussed here is the explanatory power of independent variables. One of the first sentences in the classical book 'Organizations' is: 'Authors are often convinced that the particular subjects with which they are dealing are more significant than the world has acknowledged' (March and Simon, 1958, p. 1). Analogous to this statement, one can state that by singling out one or a limited number of independent variable(-s) of a broader range, one intentionally or unintentionally creates the impression, that the selected variables provide a more important explanation than the unselected variables. Child criticizes the structural contingency theorists for concentrating on successively environmental characteristics, technology, and size. These individual factors may all play a role if the dominant coalition perceives these to be relevant and acts consequently, but it is this last step that is vital to explain structural variation. A similar phenomenon occurs in studies on the implementation of technical systems on the division of labor. By studying the changes in the labor organization coinciding with the introduction of technical systems, the impression is easily created that the technical system rather than the underlying design orientation determines the organization of labor. Here the position is taken that the impact of technical systems on the division of labor is limited, although many studies suggest otherwise.

Thus, stressing individual factors may lead to a neglect of other factors and interrelationships between factors. A full and realistic account necessitates to take into account a multitude of factors, making the total model intransparent. Obviously, this ideographic standpoint can not be maintained in research involving more than one unit, for instance in comparative research. In the underlying study, some factors are selected which are generally assumed to influence the division of labor. Those factors are selected that are thought to be most relevant for the purpose of answering the central research question about the division of labor around industrial robots. Some hypotheses are formulated concerning the relationships studied. These hypotheses function as a guideline to structure the discussions in the following.

Figure 2.2 is a graphical representation of the relationships between the aspects discussed in the preceding sections.

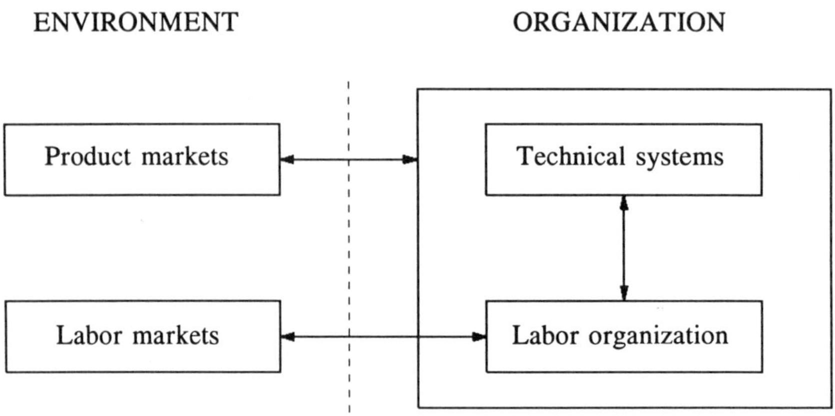

Figure 2.2: Research model

The model is a modification of a model used by the OSA, a Dutch foundation commissioning and financing research on labor studies. The model is used in many Dutch research projects as well as in the last two biannual summarizing reports of the OSA (1990 and 1992). For the purpose of this discussion Figure 2.2 shows reciprocal relationships between the distinguished elements (depicted by arrows). Rather than discussing the often unsolvable issue of the direction of the causality, the focus is on the interdependence of the elements (cfr. Sorge and Streeck, 1988). An important reason for this is the prevention of 'chicken and egg'-like questions such as 'does strategy follow structure' or is it the other way around? Probably, both are possible, the quintessence being that strategy and structure have to fit together. Structural characteristics may give rise to a new strategy but a new strategy can also be the basis for changing structural characteristics. Such a construction allows for a substantial degree of freedom, as organizations may intervene at various points. Given a certain production organization one can focus on certain markets ('strategy follows structure'), or having found an attractive market niche, one can gear its production organization to the niche. The process can be deliberate or emergent. Having selected a product market, the organization may try to influence the market conditions by measures concerning the competitive structure and\or marketing (Child, 1972).

One modification of Figure 2.2 compared with the OSA model concerns the omission of the 'production organization'. Under influence of MST, much Dutch literature makes a distinction between 'production organization' and 'labor organization'. Whereas labor organization refers to the division of labor over employees, the notion of 'production organization' has been used at various aggregation levels. The OSA uses the term for 'the way in which the activities to be performed are brought together into groups and mutually connected' (1990, p. 81) and gives as examples a line structure, a functional structure and a group structure. These same three structures are used in the COB\SER study which defines production organization as 'the way in which people and (CNC-) machines grouped c.q. coupled' (1988, p. 22). The OSA and COB\SER refer to coupling of working stations within the manufacturing department. The line and functional lay-out are typically associated with fractionized organizations, whereas the group lay-out is seen as facilitating the implementation of the integrative form of group work. Next to the grouping of machines or work stations within the manufacturing department 'production organization' has also been used to denote the division of labor over several departments (Alders and Christis, 1988b). These different aggregation levels are not always distinguished. For instance, the OSA mentions 'line organization' and 'autonomous task groups' as examples of production organization (1992, p. 85). Because of these ambiguities together with the focus in the empirical part on stand-alone working stations, the term 'production organization' will no longer be used in this study. Instead, the intra-departmental division of labor is referred to as the 'vertical dimension' of the division of labor and the interdepartmental division of labor is called the 'horizontal dimension' of the division of labor, thereby acknowledging that both dimensions are merely manifestations of the same phenomenon at different aggregation levels.

Two different kinds of the division of labor can be distinguished namely the technical division of labor and the internal division of labor. The technical division of labor is the assignment of tasks to machines respectively human beings. The term 'internal division of labor' is reserved for the (vertical and horizontal) division of labor within the organization, i.e. the assignment of tasks to individual organization members.

By distinguishing a technical and internal division of labor it is implicitly acknowledged, that technical systems influence the division of labor. The steady introduction of new technical systems ('mechanization' or 'automation'; the distinction becomes increasingly blurred with the introduction of micro-electronic control elements in traditional machines) in organizations generally aims to replace human beings by machines. This elimination of work for human beings is supplemented by the simultaneous creation of machine-related tasks. Hence, the implementation of a new technical system leads to changes in the organizational task pool, i.e. the totality of tasks that have to be performed within an organization. Whereas these changes are often incremental from the point of view of the organization, they may be substantial at the workspot(-s) affected. Its way of operation is a (mechanical) characteristic of a technical system but may pose demands on the required skills, which will be discussed in more detail below.

If the influence of the introduction of any new technical system on the division of labor is investigated, the changed organizational task pool is only one, and not even the most important part of the explanation for changes in the division of labor. More important is the internal division of labor: how are the tasks, that together make up the organizational task pool, assigned to individual employees? Deskilling, reskilling, or polarization may not be attributed to a new technical system, but are in the first place a result of assigning tasks to employees. Thus, hypothesis 1 is formulated:

H1: The application of organizational design orientations is independent from the technical system(-s) used.

Labor organizations can be divided into two polar types namely the fractionized and integrative organization, depending on the extent of the division of labor. In the fractionized organization the division of labor is maximized, while the integrative organization strives toward a minimalization of the division of labor. This characteristic has consequences for many characteristics of jobs such as the extent employees are able to, and actually perform various jobs, the number of hierarchical levels, and employee attitudes and behavior. The rules to divide the organizational task pool over organizational members can be made explicit via the application of specific design methods, but often remain implicit, although the general logic of the minimalization c.q. maximalization of the internal division of labor can be construed as manifestations of design orientations. Any organizational task pool whether affected by the implementation of a new technical system or not is divided over a number of employees via such design rules. Deskilling, reskilling and polarization can occur independently of changes in technical systems. Technical systems as such play a minor role. As argued in section 2.2, even technical systems that are explicitly developed with an integrative labor organization in mind, may be used in a fractionized organization. The same example also demonstrates that technical systems do not necessarily have to be taken as a given. Human design criteria can be taken into account, provided these can be specified. A similar but reverse process has taken place in the case of some larger production systems, in the sense of a combination of individual machines into one technically integrated production system. These may incorporate decisions with respect to the labor organization, such as machine-paced working stations in an assembly line. This may not be interpreted as technical determinism, i.e. the technical systems (directly) determines (part of) the labor organization but rather as the often implicit application of organizational design rules in a technical design.

The organization does not operate in an environmental vacuum. In sections 2.3 and 2.4 the product and labor markets were discussed. These are located respectively at the output and input side of an organization. At the labor market the organization purchases labor as its input which is a central production factor, creating output to be sold at the product markets.

At product markets, the 'flexibility paradigm' (Okko, 1992) seems to have become the dominant paradigm. This means that compared to earlier periods, the output characteristics product range, product variability, and product complexity have increased and batch sizes have decreased. The empirical basis for this trend is at best shaky, although a large number of authors describe a similar trend. Statistical and longitudinal data are lacking, which makes it impossible to assess its diffusion. Probably, large inter- and intra-branch differences occur.

The output characteristics are likely to have consequences for the division of labor at workspot level, as is more precisely formulated in hypothesis 2:

H2: The lower product complexity, the smaller product variability, the smaller the product range and the larger the batch sizes, the less likely it is, that integrated jobs are found.

In empirical research, the appropriate aggregation level of these output characteristics has to be taken into account. At the shopfloor of a 'flexible firm', the output characteristics may be ideal for a fractionized organization, if modular product design, or a similar technique, is applied (hypothesis 3).

H3: The application of variety reducing techniques such as modular product design and cellular manufacturing, change output characteristics in such a way, that favorable conditions are created for the application of fractionized organization designs.

The term 'function structure' can be seen as an equivalent of 'labor organization'. By applying design rules to the organizational task pool, tasks are grouped into functions, which together make up the function structure. This function structure forms the organization's demand for labor, which can be confronted with the supply of labor. In section 2.4, underutilization was shown to be a common phenomenon during the first half of the 1980s in the Netherlands, due to persistent high unemployment levels. The claim that the qualitative employment level has to be raised by applying integrative designs was called normative and prescriptive, as it sharply contrasts with the actual trends on the labor market. However, with a decreasing unemployment, this situation may change, as employers may see themselves forced to offer qualitatively better jobs in order to be able to attract and retain personnel. Here a distinction has to be made between a general scarcity of personnel on all labor market segments, or shortages at specific segments or on specific levels within those segments. Where scarcity exists, the scarce employees are able to select attractive jobs. The attractiveness of jobs is likely to be related to its content, either directly or indirectly via the coupling of job content to wages. At the intermediate labor market segments, scarcity may lead to the creation of functions consisting of tasks perceived to be attractive by the scarce category of personnel, leaving unattractive tasks for lower skilled personnel (hypothesis 4).

H4: Labor market scarcity at intermediate segments forms an obstacle for the application of integrative designs.

Thus, even if integrative tendencies occur, they do not necessarily lead to regradation. Underutilization is likely to disappear when unemployment resolves, but possibly only at those particular segments, where personnel shortages occur.

Furthermore, perceived uncertainty concerning future developments and a (perceived) increased need for innovation may induce employers to raise skill requirements, especially if integrative designs are used and a larger variety of tasks is combined into functions. Because of these factors and because employers seem to expect benefits from hiring over-qualified employees, the widespread underutilization can be expected to persist in the future.

Besides the product and labor markets, various other environmental parts may influence the internal division of labor. Technical and economic developments, governmental policies and legislation, the system of industrial relations, national cultures, dominant philosophies of life for instance reflected in political systems, may all to a certain extent influence organizations, and hence the internal division of labor as reflected in the organizational structure. It goes without saying that an elaborate discussion of all these factors goes beyond the scope of this study. On the other hand, some concise discussion seems necessary.

Culturalist (Hofstede, 1980 and 1990) and institutionalist (Lane, 1989; Sorge, 1991; O'Reilly, 1992) studies point to, and provide interpretations of, national differences in labor organizations. According to the Aix-school (Maurice, Sorge and Warner, 1980; Sorge, 1991) German organizations have less hierarchical levels and employ more craftsmen than comparable French and British organizations. These differences are attributed to different educational systems and career patterns, which are in use in these countries. For instance, the widely used German apprenticeship system provides a steady supply of skilled workers, whereas French organizations rely on semi-skilled and unskilled personnel. In this way, the qualitative level of labor supply influences the organization structure. Hofstede points to cultural differences between countries, defined as 'the collective programming of the mind that distinguishes the members of one category from another' (Hofstede and Bond, 1986, p. 6). At the national level, Hofstede mentions five different dimensions to categorize cultural differences, which may influence the effectiveness of particular organizational forms in certain countries. For instance, the disagreement between the American Taylor and Frenchman Fayol concerning the unity of command which was mentioned in section 2.1, is traced back to such cultural differences (Hofstede, 1990, pp. 146-148). Although such connections between psychological values dominating in a certain country and organizational practices have a certain value, institutionalists would point to different national institutions as the primary source of explanation for national differences. Within the scene set by different national frameworks, universal economic principles apply (cfr. Sorge and Warner, 1986, pp. 40-42). Unfortunately, the Netherlands have not been included in this kind of international comparative research, with one recent exception. According to Mason, Prais and van Ark (1992), who studied Dutch and British food-processing and engineering firms, the

Dutch educational system is close to the German one.

To hold the national environment constant, the pragmatic solution was chosen to concentrate the empirical work on the Netherlands. By delineating the area of research geographically, institutional and cultural factors are largely neutralized.

Finally, a note has to be made on technical developments. The other major production factor, next to labor, are technical systems. These are bought on the external market or developed internally. In either cases, factors such as the technical state-of-the-art, knowledge about the availability of technical systems, and the macro-economic diffusion of technical systems play a role in decisions of individual organizations to adopt certain machinery as discussed in chapter three.

Notes

1. Admittedly, the fractionized and integrative design orientation have much on common with existing typologies such as Burns and Stalker's (1961) 'mechanistic' and 'organic' systems of management (see Lammers, 1983, pp. 106-132, for an overview). Existing typologies, however, are often closely associated with specific theoretical approaches, in the case of Burns and Stalker, the contingency-approach. By introducing these terms, such affiliation is avoided.
2. Kuylaars (1951, p. 52 and p. 70) refers to page 59 of *Shop Management* as the place where the well-known adage, ascribed to Taylor, of the 'one best way to organize' can be found. As no such statement is made on this page, one wonders whether this statement is Taylor's. Probably, it only refers to the 'quickest time to perform a job' rather than to the structuring of complete organizations.
3. Work in groups, *Gruppenarbeit*, was promoted as early as 1922 in Germany, but was still characterized by a high division of labor (Sydow, 1985).
4. At a societal level, technological determinism generally refers to the latter meaning, while it has the first meaning at the level of organizations, although these meanings overlap.
5. The concept behind these techniques is not new: it was used for the terracotta army of the Chinese emperor Qin Shi Huang, which was assembled about 2,000 years ago (Bennett, 1992, p. 777).
6. Because of the increasing use of part-time jobs, this factor was first taken into account in the 1977 data, which led to a small adaptation.

3 Automated manufacturing

The term 'automation' is not unambiguous. It can either refer to 'automatic' or to 'automated' processes. 'Automatic' means 'having a process proceed automatically', i.e. without human intervention. This can be realized by means of a host of techniques. Nowadays, the most common control form of automatic process control involves the use of micro-electronics. Thus, in the remainder of this study 'automated' refers to the use of micro-electronics to control production processes. As all production processes need to be controlled, the potential use of automated production control is nearly limitless. Such control may apply to standard situations where process parameters need to stay within pre-set values, or to dynamic situations where the values of the parameters or even the parameters themselves are changed.

In the first section of this chapter, the abstract notion of 'technical systems' as used in the preceding chapter is made more concrete by describing the three most common forms of automated manufacturing, namely numerical controlled machines, industrial robots, and Flexible Manufacturing Machines. To gain insight into the importance of these automated manufacturing techniques, section 3.2 addresses their diffusion in the Netherlands. Section 3.3 concludes this chapter in the conventional way.

3.1 Forms of automated manufacturing

The words 'production' and 'manufacturing' are frequently used interchangeably. Yet, manufacturing is only one stage of the production process. Production includes functions such as product design, work preparation, production planning next to manufacturing operations. All these functions can be automated to a certain extent. For instance, automated production planning and control systems may or may not include a shopfloor control module, and in contrast to the calculative phases, the creative phases of the design process are generally not automated. The same can be said for manufacturing, which is the focus here. Figure 3.1 is a graphical representation of the various production activities and the terms for their automated forms. A word of warning is in place here, as there exists a considerable semantic confusion around some of these terms. 'Computer-aided' (CA) refers to the use of a controlling computer combined with equipment for, respectively, design, work (or production) preparation, logistics, and manufacturing. As used in Figure 3.1, 'CAM' refers to the linking of manufacturing, production planning and work preparation, but the same term has also been used to describe different forms of automated manufacturing. 'CIM' generally refers to the highest form of automated manufacturing, in which all production stages are linked. A requirement for CIM is that data are transferable between the different technical components.

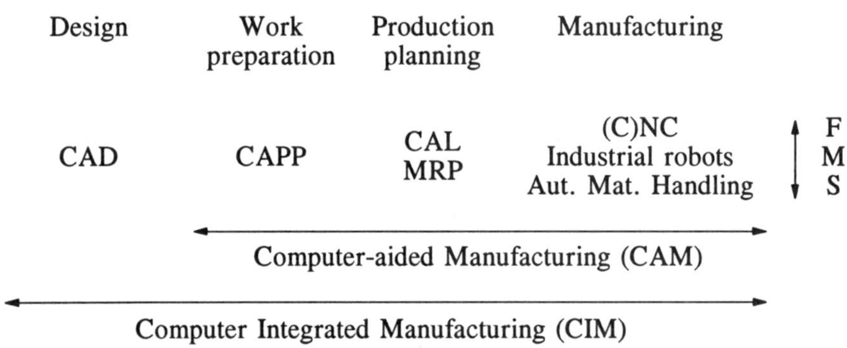

Figure 3.1: Automated production

Source: Krabbendam and Willenborg, 1986, p 424

There are many manufacturing operations involving different transformation activities, used in different manufacturing industries, and transforming varying materials, each having their own specific physical characteristics, which influence the degree to which it is technically possible to automate the manufacturing process. According to Groover (1987, p. 20), for a firm engaged in making discrete products, manufacturing consists of the following functions:

1 Processing.
2 Assembly and joining.
3 Material storage and handling.
4 Inspection and testing.
5 Control.

All functions may be automated to a certain extent. Processing and assembly are the two phases in which processes take place that are both physical and value-adding. In processing operations, the product is transformed from one stage to the next. In the transformation, some characteristic of the product is changed, such as its shape or a physical property. Processing operations can be subdivided in four categories, namely:

1 Basic processes, in which the initial form is given to the work material. Examples include casting and moulding.
2 Secondary processes, in which the final desired geometry is given to the product. Examples are turning and pressworking operations.
3 Operations to enhance physical properties, for instance heat-treating and preshrinking.
4 Finishing operations, such as polishing, painting, and zincing.

In assembly processes, two or more parts are joined together either by mechanical fastening operations (assembly) or joining processes such as welding, glueing, and soldering.

The various kinds of operations can be automated by means of (C)NC-machines and/or industrial robots. The latter are predominantly being put to use in assembly, joining, and handling, whereas the former typically concern processing.

NC, CNC and DNC

Numerical control (NC), which was first introduced in the 1950s, as an alternative to 'Record Playback' (see section 2.2).

> NC is a form of programmable automation in which the processing of equipment is controlled by means of numbers, letters, and other symbols. The numbers, letters, and symbols are coded in appropriate format to define a program of instructions for a particular workpart or job. When the job changes, the program of instructions is changed.

> [...] A numerical control system consists of the following three basic components:
>
> 1 Program of instructions.
> 2 Machine control unit.
> 3 Processing equipment.
>
> [...] The program is fed into the control unit, which directs the processing equipment accordingly' (Groover, 1987, pp. 199-200).

Typical NC-machines are controlled by means of punched tapes, which were later replaced by more advanced data storage devices. If the machine control unit consists of a computer (generally a PC), the numerical control system is computerized: 'CNC-systems', which were first introduced in the mid 1970s. Thus, the difference between NC- and CNC-machines concerns only the control system. As practically all control systems are CNC-systems nowadays, the distinction has lost its meaning, and often 'NC' is used to denote both. A direct numerical control system, or 'DNC', uses a central computer to direct the operations of a number of separate NC-machines.

The most common application of NC is controlling machine tools, which can be used for a variety of operations, such as drilling, milling, turning, shaping, and grinding. These operations can be described by means of parameters (the speed, the depth, and the feed), which makes it possible to control the machining conditions. The program contains these parameters. Other applications of NC include cutting, bending, wrapping wire, inserting components, and measuring coordinates.

An important advantage of NC-machines above conventionally controlled machines, is that several operations can be performed in one set-up in so-called 'machining centres'. Machining centres are NC-machines with some special features. They typically have the capability to change tools automatically, automatic workpart positioning, and a pallet shuttle, which allows for continuous operation. Finally, NC-machines make some complex machining operations possible, which could previously not be performed.

CNC-programming methods can be divided into three categories (de Korte, 1991):

1 Universal programming methods.
2 Machine-specific programming methods.
3 MCU-programming (MCU = Machine Control Unit).

Universal programming methods make processing programs independently from the specific processing machine, which results in a so-called 'Cutter Location File' (CL-file). The machine-specific program is generated from this CL-file by means of a postprocessor. The major advantage is that the same CL-file can be used for various machines, which is not possible using machine-specific programming methods. Universal programming methods make a link with CAD-systems possible.

Both universal and machine-specific programming methods make use of a programming computer, external to the CNC-machine. In case of machine-specific methods, the programming system used on this external computer has to be identical to the programming system used on the actual CNC-machine. With MCU-programming, the CNC-machine's controlling computer itself is used to write programs. The difference between machine-specific and MCU-programming methods thus concerns the location of the programming computer, which is respectively external to the machine or internal.

Besides these programming methods, some other terms need clarification. Shopfloor programming, also called 'machine programming', means writing programs at the CNC-machine itself, for which especially MCU-programming is suitable. When computerized control systems were first introduced in the mid 1970s, they enabled to edit, and later to write programs by means of shopfloor programming. However, this had the important disadvantage of machine downtime, as the machine's MCU had to be used for programming. When parallel programming, which allows programming during machine operation, became available, the solution for this problem was found. Nowadays, parallel programming is often a standard feature of NC-machines.

Industrial robots

The confusion around industrial robots is substantial. Robot pioneer Joseph Engelberger is quoted by Reijers and de Haas as follows: 'I cannot define a robot, but I can tell you for sure when I see one' (1986). Various definitions exist in several countries. An important difference between these definitions concerns the number of freely programmable axes, or 'degrees of freedom'. The definition of the Japanese Industrial Robot Association JIRA in particular contrasts with Western definitions. The former definition acknowledges manipulators with two degrees of freedom as robots, while the latter are more restrictive with respect to the degrees of freedom a robot must have. The International Standards Organization ISO defines 'manipulating industrial robot' as follows:

> An automatically controlled, reprogrammable, multi-purpose, manipulative machine with several degrees of freedom, which may be either fixed in place or mobile for use in industrial application (ISO/TR 8373, in: Karlsson, 1991, p. 2).

This definition was also adopted by the International Federation of Robotics IFR. Although it does not specify the number of degrees of freedom necessary to be acknowledged as a robot, there seems to be a communis opinio in the Western world that this number should be at least four. International comparisons of diffusion rates (Karlsson, 1991; Tani, 1992; Vickery and Campbell, 1989) may show some distortions because of this lack of consensus.

Robots can be used for a variety of operations, both in manufacturing and other industries. Worldwide, the most popular applications of industrial robots include arc welding, spot welding, parts and tool handling, and assembly (Karlsson, 1991, pp. 151-154). A new area of development concerns micro-robotics, which are often used for assembly operations. Non-industrial applications of robots include agricultural uses, cleaning of nuclear waste, detecting bombs and underwater robots.

The general public, fed by articles in the popular press, seems to hold an image of the robot as a supermachine, fitted to produce flexibly a large range of products and perform many different operations. Various elements of the ISO definition confirm this image of 'flexibility', namely 'reprogrammable', 'multi-purpose', and 'several degrees of freedom'. However, as will be argued in the following, the use of robots is far more restricted than the definition suggests. Two elements of the ISO definition deserve special attention in this respect, namely:

- the multifunctionality ('multi-purpose') of industrial robots, which is determined by the robot's body and its effectors.
- the reprogrammability of industrial robots, which is determined by the used programming method.

The degrees of freedom refer to the number of axes. Most robot bodies consist of three freely programmable axes, which can be moved via translations or rotations. The combination of translational and rotational axes is called 'kinematical configuration', and can be denoted by capital letters; 'RTT' stands for the combination of one rotation and two translations. In 1988, the RRR-configuration was the most popular configuration in the Netherlands. As far as the quality of the dataset allowed to draw conclusions, 85 per cent of all Dutch industrial robots in 1988 had an

RRR configuration. The RRR-configuration is used for at least eight different applications (Benders, 1991a, p. 803). Such broad variety of applications might have helped to give the robot an aura of multifunctionality. However, the body of a robot needs to be supplemented by effectors, which are grippers or tools to perform the specific operations the robot is purchased for. These effectors often form a substantial part of the investment in robots, as they frequently have to be designed and made according to customer specifications. It is not unusual that effectors and other peripheral equipment make up two thirds of the total investment in a robot system. Given the fact that effectors are always specific for some application and often for a specific production process, the multifunctionality of industrial robots is limited by the extent to which the effectors are specific. Furthermore, switching effectors causes a loss of production time. Finally, robots are sometimes used as part of a larger technical system, within which they perform some specialized task (often parts handling). Thus, hardly any robot is used for more than one particular application. Robots which are included in larger technical systems may form exceptions: some of these perform two or more operations one after the other, and thus have to switch tools repeatedly. However, most robots have to be characterized as dedicated machines. Many robot producers are even specializing on certain applications such as arc welding.

Although many industrial robots are controlled by means of NC, the way of programming robots is generally different from the programming methods used for CNC-machines. An important difference between robots and CNC-machines concerns their geometry. Whereas machining tools and workpieces are often placed in relatively fixed positions, which can be easily mathematically modelled and controlled, the complex geometry of industrial robots makes the necessary calculations substantially more difficult, especially if many rotating axes are used. There are several (computerized) programming methods for industrial robots:

1 Leadthrough programming or 'teach-in', which has two variants:
   a. manual leadthrough
   b. powered leadthrough.
2 Textual language programming or computer-like programming methods.
3 Off-line programming.

Leadthrough programming requires the programmer to move the robot arm through the desired motion path, which can either be done manually or via a 'teach pendant'. Manual leadthrough is only occasionally used. It is convenient to describe irregular motion patterns such as in paint spraying. Powered leadthrough is probably the most common programming method. It is suited for point-to-point control, as used in arc welding. A teach pendant is a hand-held control box, which has toggle switches or contact buttons for controlling the movements of the robot arm. The operator drives the robot arm into the desired positions, which are recorded in the memory. During subsequent playback, the robot moves through this sequence of position on its own power. This form of robot programming can only be done on-line, which may lead to considerable downtime of the robot. Therefore, this programming method can only profitably used when substantial numbers of a specific product have to be produced, i.e. long production runs and batches (Groover, 1987, pp. 320-325).

Textual language programming is fairly similar to CNC-programming. A number of abstract instructions, 'textual statements', make up a computer program. For instance, 'MOVE X2' means that the robot has to move to point X2. Advantages of this programming method include the possibility to use enhanced sensor capabilities, improved logic control, and possibilities to communicate with other computer systems. In motion programming, textual statements are used to describe the movement, but leadthrough techniques are used to define the position and orientation of the robot. The statement 'HERE X2' then means, that the reference point X2 is fed into the robot's control system. This combination of programming methods led to the name 'hybrid programming methods'. Input signals from sensors can be used in so-called 'interlock commands', which may start or end robot movements.

Pure textual language programming can be seen as one form of off-line programming. Perhaps, off-line programming can be defined best as 'a way of programming in which ready-to-use programs are written on an external computer'. The element 'ready-to-use' means, that it is no longer necessary to adjust the program to the actual robot set-up as is the case with hybrid programming. Although development work on off-line robot programming languages started in the late 1970s, they are scarcely used, mainly because of the substantial investments in an off-line programming package and 'calibration' difficulties. Calibration is the process of adjusting for geometrical differences between the model used by the off-line system and the actual robot cell, and forms a major obstacle for the use of off-line programming (MacConaill, Drews and Robrock, 1991).

Compared with on-line programming methods, off-line programming has the advantage that programming and regular production can proceed simultaneously. Thus, programming does not lead to a loss of production time. A second, probably more important reason to use off-line programming, is the simulation of robotized cells. Graphical off-line packages allow to simulate these cells in order to optimize their design. Robot producers and suppliers may use this option. However, a graphical module is not necessarily part of an off-line programming system: a spreadsheet may suffice. Furthermore, their use may become necessary if robots are integrated into larger production systems (data transfer), and if there are high precision requirements (Vermeulen, 1985, p. 408).

Although Gorbachev et al. (1991, p. 24) mention over 90 known systems, few cases of actual implementations are known (MacConaill, Drews and Robrock, 1991). Assembly is the main application area, although welding becomes increasingly important (cfr. Hollingum, 1991). In the Netherlands, the (non-graphical) package ARAC was developed for cutting and welding purposes.

According to Scherff (1989), off-line programming suffers from the disadvantage that it is not done at the actual workspot. She argues that the necessary knowledge to work with graphical techniques is generally not present on the shop floor. This makes the programs susceptible to programming faults, as vital operator knowledge can not be used. Therefore, the Aachen, Germany, based RWTH developed a speech-driven programming method, which allows operators to compose programs.

Another development concerns the possibility of parallel programming, which some robot suppliers offer. Parallel programming also allows programming during the operation of the robot.

Flexible Manufacturing Systems

Early developments on Flexible Manufacturing Systems took place around 1970. Flexible Manufacturing Systems exist. Especially in the first half of the 1980s several forms of Flexible Manufacturing configurations were distinguished (FM units, modules, cells, systems, factories), but only the distinction between FM Cells and FM Systems seems to have lasted. By comparing various definitions (Verbraeken, 1982, p. 40; Bilderbeek and Kalff, 1985, pp. 40-46; Economic Commission for Europe, 1986, pp. 13-19; Groover, 1987, p. 464), the following list of minimum requirements to be classified as a FMS was compelled:

- it should consist of at least four machines;
- between which workpieces are transported automatically;
- which are centrally computer controlled (generally by a DNC);
- and have a tool magazine, a pallet magazine, as well as automatic

tool and pallet changing devices;
- load and unload stations;
- workpiece routing is automated.

Additional features may include:
- automatic tool inspection;
- automatic workpiece inspection;
- storage of half and end products.

Less complex configurations are labelled 'FMC', whereby the number of machines is seen as essential: FMCs consist of a maximum of three machines. Industrial robots can be incorporated into a FMC\FMS to perform functions such as material handling and tool switching. Groover (1987, p. 464) makes a distinction between dedicated and random-order FMSs. Dedicated FMSs are designed to produce a limited variety of workpieces, while random-order FMSs can handle substantial variations in the number of workpieces. The latter system is more advanced; it needs more tools and a more sophisticated computer control system, and is therefor more expensive.

3.2 The diffusion of automated manufacturing in the Netherlands

The most reliable and complete data concerning the diffusion of manufacturing automation in the Netherlands are provided by the Dutch Central Bureau for Statistics, CBS. The information in this section, which is not reported in these publications, was gathered by a telephonic interview with the responsible CBS researcher. The CBS data are occasionally supplemented with data from other sources, if these provide additional insight.

Before presenting the CBS data, attention has to be paid to the CBS definitions c.q descriptions. Computer Aided Manufacturing CAM is described as 'all use of computer controlled machinery for manufacturing products, which can range from discrete semi-manufactured articles to end products (like furniture) on the one hand to non-discrete ('flow') products, such as oil-based and pharmaceutical products' on the other hand (CBS, 1991, p. 39). Data are also presented for (C)NC-machines, DNC-machines and industrial robots. DNC-machines are (C)NC-machines which are linked to each other and controlled from one central point (CBS, 1991, p. 44). Flexible Manufacturing Systems are not included as a separate category. Concerning industrial robots the following statement is made: 'Although in the questionnaire robots are considered to be handling appliances that have two or more freely programmable axes, further research showed that the robots included have at least three degrees of freedom' (CBS, 1991, p. 44), which refers to a reliability check after the initial data had been gathered.

Tables 3.1 and 3.2 contain the absolute numbers of CNC-machines, DNC-machines and industrial robots. The data for 1986 and 1990 are prognoses. In 1986 and 1990, firms were asked to give their investment plans concerning the forms distinguished. The 1986 data exclude the first-time users, the 1990 data include this category.

Table 3.1
CNC-machines and DNC-machines in Dutch manufacturing, 1988-1990 (absolute numbers)

Machine	Branch	1988	1990
CNC	Food processing	1724	2086
	Textile \ leather	281	454
	Paper \ graphics	1191	2373
	Metallurgic	5647	7566
	Other branches	931	1310
	Total	9774	13789
DNC	Food processing	3167	3966
	Textile \ leather	153	375
	Paper \ graphics	227	298
	Metallurgic	941	1654
	Other branches	1962	4719
	Total	4770	6736

Source: CBS, 1991, p 45

Table 3.2
Industrial robots in Dutch manufacturing, 1984-1990
(absolute numbers)

Branch	1984	1986	1988	1990
Food processing	15	19	10	56
Textile \ leather	7	18	0	12
Paper \ graphics	34	36	73	161
Metallurgic	335	636	753	1043
Other branches	53	83	38	69
Total	444	792	874	1341

Sources: CBS, 1987, p 33 and 1991, p 46

Table 3.3 provides data on the relative diffusion of automated manufacturing in a number of industrial branches, and makes it possible to relate the data to the sizes of these branches.

Table 3.3
CAM in Dutch manufacturing, 1984-1988 (user firms as a percentage of the total number of firms per branch)

Branch	1984	1986	1988
Food processing	11	17	17
Textile \ leather	9	17	17
Paper \ graphics	18	22	28
Metallurgic	16	25	29
Other branches	12	21	25

Sources: CBS, 1987, p 25 and 1991, p 40

The CBS data were collected by means of a large, representative survey (sample of one third of the population; response rate of 67 per cent). With respect to some items, a small numbers' problem occurs.

From Tables 3.1, 3.2 and 3.3, several conclusions can be drawn with respect to the diffusion rates of the forms distinguished in the various industrial branches, and their development from 1984 onward. Firstly, (C)NC-machines are used most widely, followed by DNC-machines and industrial robots. Especially industrial robots are far less common than (C)NC- and DNC-machines. Secondly, from 1984 onwards, the use of all forms has increased in probably all branches. The absence of this result for the textiles and leather industries, and the food processing industry (Tables 3.2 and 3.3) may be attributable to the small size of their representation in the CBS samples, machines that are no longer being used, and the non-realization of the (possibly too optimistic) expectations about future investments. Thirdly, the diffusion rates of the investigated forms vary considerably over several branches of industry. The metallurgic industries are the main users of CNC-machines and industrial robots; more than three quarters of all robots are applied in metallurgic industries. DNC-machines are used more widely in food processing and 'other' industries.

These differences can be explained by various factors (Bilderbeek and Kalff, 1985). Firstly, the state of the art of automating manufacturing processes varies, which is related to the character of the operations to be automated as well as the physical properties of the materials used. For instance, assembly operations are relatively hard to automate. The same holds for materials, that have non-homogeneous properties, such as leather and timber (Bilderbeek and Kalff, 1985). The processing characteristics of homogeneous materials such as metal are (relatively) easy to model mathematically, which is a precondition for writing programs. Secondly, the average organization size in these branches varies, which is related to factors as the availability of financial resources and possibly to output characteristics (see below for a more elaborate discussion of these factors). Branches with many large firms, i.e. a high degree of concentration, are likely to have higher levels of automated machinery.

According to Hagedoorn, 'Sophisticated systems of automated manufacturing such as Computer Integrated Manufacturing Systems are reserved for the very large and large companies. Lower levels of automation and incremental improvements are more appropriate for small and medium-sized companies' (1988, p. 49). Table 3.4 contains data about the use of automated manufacturing in firms of different sizes (measured in number of employees).

Table 3.4
Firm size and use of automated manufacturing, the Netherlands, 1988 (user firms as a percentage of the total number of firms per size category)

Organization size (number of employees)	CNC	DNC	Robots
5 - 20	15.6	2.5	1.7
20 - 100	27.5	4.7	2.6
100 - more	40.9	12.9	8.2

Source: Calculated from CBS, 1991, pp 40 and 45-46

From Table 3.4, it follows that the firm size is positively related to the use of automated manufacturing, which holds for all three forms, and is confirmed in Poutsma (1991, p. 44). Especially DNC-machines and industrial robots are used more widely in the largest size category. For DNC-machines, this may be due to the situation in the process industries as described above. Concerning the use of industrial robots, it has to be noted that the presence of two large users, namely the NedCar (former Volvo) automobile plant in Born, which includes a spot welding line (also see section 4.3), and Philips (several large assembly plants), probably increases the average substantially. As both the DNC-machines in processing industries, and many industrial robots in the NedCar and Philips plants are probably part of larger automated systems, this can be read as a confirmation of Hagedoorn's statement, that large firms are more likely to use sophisticated systems, while small and medium-sized firms tend to use stand alone equipment.

There are several explanations for the larger use of automated manufacturing in large firms (Bilderbeek and Kalff, 1985):

Large firms can expect to possess or have access to larger financial resources. As prices of the several appliances gradually drop, and gradual advances in R&D and technical knowledge are made, this argument can be expected to loose weight over time.

The risk of investing in automated manufacturing appliances may very well be larger for small firms. As a proportion of total investments,

investments in automated manufacturing are often substantial, as many projects are tailored to customer needs. The effects of implementing such new production techniques on the entire organization may be perceived to be more drastic than in larger organizations.

The level of knowledge, which is a key factor leading to the adoption of new production techniques (Rogers, 1962, p. 13; Schünemann and Bruns, 1986), about automated manufacturing is often quite low in small organizations (Docter and Stokman, 1987; Stokman, de Groot and Leurs, 1989, p. 77; KNOV, 1991, p. 47). Docter and Stokman (1987, p. 94) report, that Dutch small and medium-sized manufacturing organizations implementing process innovations rely on suppliers as the most important source of information. According to Stokman, de Groot and Leurs report (1989, pp. 77-81) these enterprises tend to make little use of independent sources of information. Instead, they often rely on (free) information from suppliers. In december 1988 the first so-called 'Innovation Centres' were founded to encourage the transfer of knowledge from knowledge holders, such as consultancy agencies in engineering and universities, to small and medium-sized enterprises. These Innovation Centres act as a kind of information brokers.

In contrast to their image, 'flexible' manufacturing techniques are not always suited to handle a large product variety. Roughly: the greater the variety that can be handled by automated equipment, the more expensive it is. The handling of a substantial number of products by automated equipment may be necessary to keep this production economically feasible (see section 3.1). The larger part of small and medium sized enterprises in manufacturing industries produces small batches of ever varying products, for which flexible manufacturing automation is not suited (Stokman, de Groot and Leurs, 1989, p. 100).

Table 3.5 deals with regional differences of the diffusion of automated manufacturing in the Netherlands. The data are retrieved from a number of reports by the main employers' organization for the metallurgic industries, FME. Medio the research period (third quarter of 1989), the metallurgic branches had 4090 firms with a minimum of 10 employees (CBS, 1990, pp. 12-13); the FME had 'about 1260' members. The FME data are not representative for the Dutch metallurgic industry as a whole. The investigated firms have an average workforce of 225, confronted with 75 in the metallurgic industries as a whole. As larger firms tend to make more use of manufacturing automation, FME-members can be expected to make more use of automated manufacturing than the average firm in the industry. This is actually confirmed in a confrontation of Tables 3.3

and 3.4: the diffusion of automated manufacturing among FME-members has gone further than in the metallurgic industry in general. However, assuming that the characteristics of FME-members in the several regions are fairly similar, the FME data give an adequate picture of regional differences in the diffusion of automated manufacturing in the Netherlands.

Table 3.5
Regional differences in the diffusion of automated manufacturing in the Netherlands (percentage of member firms using application)

Region	N	Date report	CNC	Robots	PLC	Future plans
East	113	1 - 1989	47	23	37	87
South	122	4 - 1989	57	16	31	96
North Holland	106	7 - 1989	47	10	24	79
Rhine delta	105	10 - 1989	32	3	11	73
North	109	1 - 1990	28	8	16	86
Middle West	110	3 - 1990	32	4	14	81
National	665	8 - 1989	41	11	22	84

Sources: FME, 1989a-b-c-d and 1990a-b-c

From Table 3.5 it can be concluded, that the northern and central regions lag behind the South (a traditional stronghold of metallurgic industry) and East. North Holland, including Amsterdam, holds an intermediate position, scoring about the national average on all three forms. Judging from the plans concerning future investments, the forerunner role of Southern and Eastern regions is likely to continue in the near future, provided of course, that these plans are materialized. The differences are caused by regional concentrations of specific industrial branches, as well as differences with respect to 'innovativeness' in these regions. Considering the existence of such regional differences, one wonders whether the national level is always appropriate in comparative research between countries. Figure 3.2 describes the diffusion of industrial robots in the Netherlands.

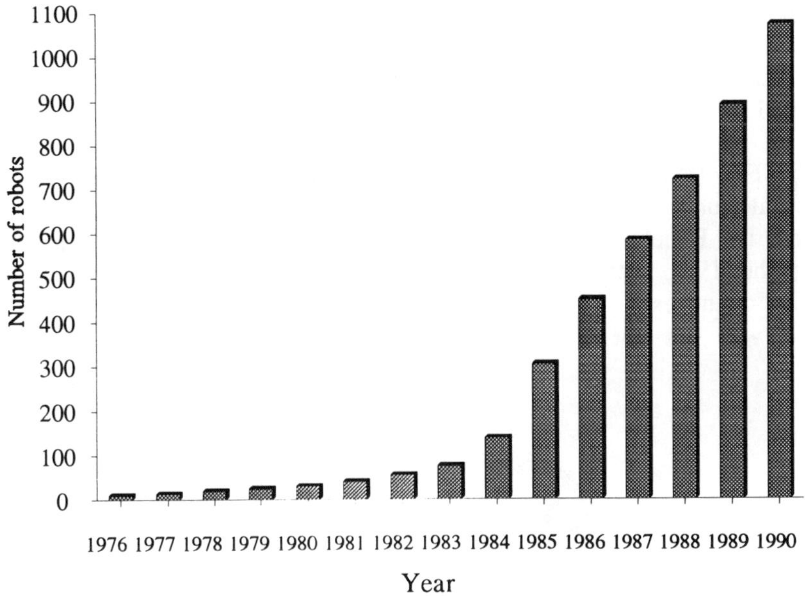

Figure 3.2: **Diffusion of industrial robots in the Netherlands 1976-1990**

Sources: Vermeulen and van Veen, 1991 and ANERTEK, 1989

The data are derived from two reports published by the consultancy agency ANERTEK (1989; Vermeulen and van Veen, 1991). Their basic data were acquired from robot suppliers, who provided lists of their customers. The 1989 data suffer from some limitations:

- two suppliers, among whom market leader ABB, did not participate;
- fourteen users did not allow the publication of their addresses, but were included in the aggregate data;
- only Dutch suppliers are included;
- internally developed robots are not included.

The 1991 report does not contain such detailed methodological information. However, it is stated that a growing number of robot suppliers refused to cooperate as a previous report contained some statements which

were interpreted negatively (Vermeulen and van Veen, 1991, p. 9). However, these data are the most detailed known to be available.
Figure 3.2 clearly shows that the diffusion got an upswing in the second half of the 1980s. From 1984 onward, the annual increase is between 134 robots (1986) and 181 robots (1990).

ANERTEK (1989) also provides data concerning the applications of industrial robots, which are shown in Figure 3.3 (the 1991 survey covered the Benelux instead of only the Netherlands as was the case in 1989; the 1991 data about applications are not split up between the separate countries).

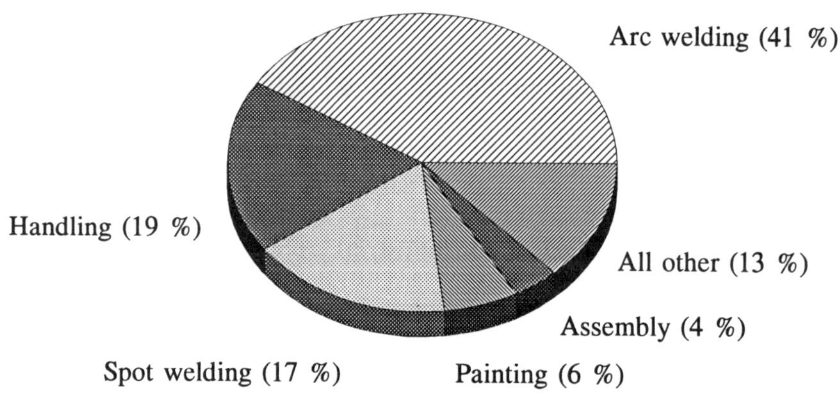

Figure 3.3: Applications of industrial robots in the Netherlands in 1988

Source: ANERTEK, 1989

Except for a few, all spot welding robots were used in NedCar's (formerly Volvo) Born plant. Although assembly seems to be underrepresented, the exclusion of internally developed robots may distort the image. Especially Philips uses many machines in assembly operations, which according to the definition of industrial robots have to be considered as such, but which are seen as specialized machines. Philips uses approximately 200 IC-bonding machines in its Nijmegen plant, which satisfy the requirements of the definition of industrial robot. However, at Philips such machines are no longer considered to be a separate category, and no accounts are available concerning their use. Hence, the conclusion is warranted that there are probably more assembly robots than the ANERTEK data show.

Arc welding is by far the most popular application, as 41 per cent of all robots are used for arc welding. The percentage of spot welding robots dropped from 25 per cent in 1987 to 20 per cent in 1989. As practically all spot welding robots are used in the NedCar plant, this fact points to the acceptance of this 'new' technology in small and medium-sized enterprises.

Compared with the data presented by Laurentius, Timmerman and Vermeulen (1982, pp. 5-7) and the CBS data, the ANERTEK data are lower, despite the fact, that 83 robots in instructional institutions are included in the ANERTEK data. This can be explained by the above mentioned limitations in the ANERTEK data, and because the CBS guarantees anonymity, which raises the participation to cooperate. Both reports show that industrial robots are increasingly being used, be it that their diffusion may proceed slower than expected. In 1986, Reijers and de Haas expected the number of robots to double every three years. 'It can be expected that around 1990, the penetration of industrial robots will be such, that they will form a significant economic factor' (Reijers and de Haas, 1986, p. 12). The data in Tables 3.2, 3.4 and 3.5 and in Figure 3.2 show that both statements, especially the last one, turned out to be overestimations, and can be interpreted as a prognosis, which turned out to be too optimistic. Later, Reijers confirmed that the diffusion of industrial robots proceeded more slowly than was expected in the beginning of the 1980s (1992, pp. 43-44).

In 1985, Bilderbeek and Kalff reported that no flexible manufacturing systems were known to be operational in the Netherlands. The Economic Commission for Europe of the United Nations (1986, p. 25) mentions a total of 72 FM Cells and 25 FM Systems, installed in the Netherlands in 1984/85. The consultancy agency Berenschot estimates that the number of Dutch FMCs\FMSs in 1987 was approximately 20 (1989, p. 13), and the IIASA-database contains 8 Dutch FMSs (Tchijov, 1992, p. 212). The only conclusion that can be drawn is, that there is a considerable degree of disagreement concerning the diffusion of FMSs in the Netherlands.

Slomp et al. (1989 and 1990) conducted a study about possibilities for future placements of FMS's in the Dutch machine building companies having 50 employees or more, and concluded, that some 35 per cent of metal cutting companies in the Netherlands are potential FMS-users. The following factors were identified as the four most important bottlenecks for the implementation of FMSs:

- experience with production automation.
- good financial results.
- attention for strategic planning.
- the absence of shift systems.

Concerning the highest level of manufacturing automation to be achieved, Computer Integrated Manufacturing, Vrakking mentions a project at Holec in Hengelo, which is 'unique in the Netherlands' (1988, p. 110). It consists of a combination of computer aided design, logistics, and various production techniques (CNC, DNC, robotics, automated material handling). At the time of the report, the Holec project has been partially realized, be it with delay.

3.3 Conclusions and discussion

CNC-machines, industrial robots, and Flexible Manufacturing Systems are all advanced manufacturing techniques, which will ultimately become ordinary standard machines. However, their diffusion rates vary considerably, as demonstrated in section 3.2. Probably, similar conclusions can be drawn for most industrialized countries. Whereas CNC-machines are a general phenomenon in the 1990s, industrial robots are still quite rare despite their broad range of applications for which they are fit, and Flexible Manufacturing Cells and Systems can still be seen as proof of an organization's outstanding innovativeness. The more complex these techniques are, the less they are applied. In the case of these advanced manufacturing techniques, 'complexity' refers to technical difficulties that have to be solved in order to become a standard and economically feasible alternative to other ways of manufacturing. The kinematical configuration of industrial robots has consequences for the programming methods used, which in its turn is a barrier for their diffusion. FMSs are composed of many advanced technical components that need to be balanced, which is a complex technical issue. Solving these issues requires substantial financial resources, which is an obstacle for large-scale use of these techniques. These technical constraints limit the width of the product range that the equipment can deal with. The more complex a system is, the more expensive it is, the less attractive it is as an economic alternative, unless this disadvantage is neutralized by an increased machine productivity, i.e. an increasing number of products that is being processed during a certain period. Combining different products to create a sufficient total volume is one solution, provided this is technically and economically feasible. For CNC-systems this is roughly the case, but robots and FMSs incur problems in this respect. The widely used teach-in programming method is an obstacle for handling a large range of products in case of robotics. Furthermore, an argument that is valid for all three forms is that different products may require different fixtures, which generally require substantial investments. In case of FMSs, all individual system components need to be able to handle the various operations needed to process varying products individually, and the components have to function as a collective system too, which requires the handling of a complex process. The technical difficulties are gradually being solved, which in combination with increasing knowledge and decreasing prices, is likely to lead to an increasing adoption of these technical systems.

Besides the mentioned techno-economic barriers, several other arguments were mentioned in section 3.2 as bottlenecks. The distribution of knowledge to small and medium-sized firms and\or existing experience with production automation, (lack of) financial resources either because of unfavorable economic results in the future or inadequate access to external sources, and the Dutch conservative attitude with respect to working hours are some of these barriers. The persistence of such barriers provides one explanation for the (for technicians undoubtedly) disappointing outcomes of earlier expectations concerning future diffusion rates, as mentioned repeatedly in section 3.2.

An important issue concerns the development of various programming methods. In section 3.1, these were described for CNC-machines and robots, which makes it possible to make a comparison.

Before the advent of parallel programming, MCU-programming was resisted as the machine's computer could not be used at the same time for manufacturing operations. For CNC-machines, the issue was in a technical perspective resolved with the advent of parallel programming, but the same problem still exists for robots. Teach-in programming, which is commonly used for industrial robots, incurs downtime. Off-line programming methods are being developed and gradually introduced to overcome this problem. The second alternative to prevent downtime, namely parallel programming, has only recently become available for some robots. Thus, the teach-in programming method has been and still is an obstacle to use industrial robots for a large product range, which problem may be resolved by off-line programming. Whereas in MCU-programming, teach-in programming and parallel programming the machine's controlling computer is used to program, this is no longer necessary if an external computer can be used for this purpose. Both for CNC-machines and industrial robots, universal and machine-specific (off-line) programming languages exist, but the necessary calibration is still problematic in case of robots. Thus, the complex kinematical configuration of robots has slowed down their degree of adoption, as the unsolved programming issue prohibits the use of robots for producing a wide product range. Via these consequences for the output characteristics, the teach-in method may also have consequences for the division of labor, as the chance to find integrated functions decreases in more stable production environments (see hypothesis 2). On the other hand, the simple teach-in programming method is easier to learn than off-line programming. The latter may be too difficult for robot operators, in other words: the skill requirements needed for respectively operating and programming may diverge that much, that both (complexes of) tasks are assigned to different employees (cfr. section 2.5 and Benders, 1991b). In chapters 4 and 5, an attempt is made to answer such questions empirically.

Note

1. By international standards, other size categories might be more appropriate. In the Netherlands, small and medium-sized enterprises are considered to have less than 100 employees (Stokman, de Groot and Leurs, 1989, p. 5).

4 Work design and manufacturing automation in the Netherlands

In this chapter, an overview and analysis of existing empirical studies on automated manufacturing in discrete processes in the Netherlands is given. Both case-studies and surveys on the division of labor around subsequently CNC-machines, industrial robots, and flexible manufacturing systems are dealt with, for as far as they present data concerning the division of labor and influencing factors as identified in chapter two. The focus is on the micro-level, namely the workspot. With respect to the division of labor, a distinction is made between the following machine-related tasks (see Kern and Schumann, 1984, pp. 184-195; Alders, Christis and Bilderbeek, 1988, p. 11):

1 programming;
2 testing and optimizing;
3 operating tasks.

If these tasks are combined into one function, the function is 'integrated'. If testing\optimizing as well as operating tasks are part of the operator's function, 'partially integrated' is used. If the operator's function does not contain testing\optimizing or programming, but just operating tasks such as (un-)loading and monitoring, the function is 'split'. These options will be termed respectively 'I', 'PI' and 'S'. Tasks which are not directly related to the specific form of automation discussed, are generally abstracted from, unless the sources used pay explicit attention to them. Furthermore, data on the output characteristics, the technical systems, the labor market, organizational size, the number of programmable machines, and (organizational) design orientation are presented as far as the material allows.

The material on CNC-machines is most elaborate; the coverage of industrial robots is limited, which is also true, but to a lesser extent, for FMSs. Some secondary studies (in the sense of only indirectly containing information about the division of labor) is therefore included in the section on FMSs.

It needs to be pointed out, that many of these studies, especially those on CNC-machines, were conducted by researchers closely associated to modern sociotechnology. The interpretation of the material is frequently and openly flavored by preferences, stemming either from MST or more humanistically motivated concerns for the quality of working life. In the following, an attempt is made to give a more objective, or perhaps objectivistic, presentation of these data. When possible, the original material is used. The sections 4.1 to 4.3 are closed with conclusions and a discussion of the results after the empirical studies have been described. Section 4.4 contains the final conclusions of this chapter.

4.1 CNC-machines

From 1985 on, several studies on CNC-machines have been conducted in the Netherlands. Peters (1986), the COB\SER-study (1988), and the EIM-study (Poutsma, van der Staal and Zwaard, 1986) contain case-studies. The TOWES-project (Alders, Christis and Bilderbeek, 1988) consists of a survey, while the EIM-study contains survey data as well. Most of these researchers are affiliated to MST. Alders, Christis, Fruytier, ten Have, and Peters are or have been members of the MST-oriented POST-group (see their 1991 book), which influenced their empirical research (cfr. Poutsma, 1991).

Peters' unpublished master's thesis (1986) contains six detailed case descriptions. The data were gathered in 1985 and 1986. Four cases are in the machine-building industry, one in transport, and one in the metallurgic branch. The original case material is analyzed in the following, focusing on the question why five of the companies decided not to decentralize programming. The sixth firm was included only after Peters had found out that it had opted for integrated functions. The study contains detailed information concerning output characteristics and motivations why particular options were chosen. Table 4.1 describes some basic case characteristics. Where more than one option occurs within one organization, these options are split by ';'. In case 4 , 'PI-I' means, that both a programmer and a CNC-operator program, but the latter makes more simple programs.

Table 4.1
Selected characteristics of six CNC-cases

Case	1	2	3	4	5	6
Number of CNCs	3	3; 1	2; 1	2; 1	4	8; 2
Option	PI	S; PI	PI	PI; PI-I	PI	I; PI
Repeat frequency	High	High	High	Low	High	0
Programming frequency	Low	Low; high	Low	High	Low	High
Parallel programming	No	No	No; yes	No	No	Yes

Source: Peters, 1986, pp 67-98

Several conclusions can be drawn from Table 4.1. Except for case 6, which was explicitly included for having integrated functions, integrated functions do not occur. On the other hand, testing and optimizing are generally part of the operator's jobs, be it that Peters reports that these are sometimes done in cooperation with the programmer. The PI-option is clearly dominant. Despite the fact, that Peters reports about integrative moves in cases 1 (project 'quality of working life'), 3 (group lay-out, elimination of hierarchical levels), 4 (strive toward multi-functional operators, although quality of working life is considered to be a luxury) and 5 (elimination of hierarchical levels), none of these organizations has established integrated functions. However, in case 6, production is organized in autonomous groups, which coincide with the dominance of the integrated option. The number of machines does not seem to play a role in these data. In contrast to what might be expected, the concentration of the programming task into a separate function does not occur in case 6.

Peters made some detailed observations with respect to output characteristics. In case 1, high product complexity is mentioned as the prime reason why operators should not program: high product complexity asks for considerable programming skills, and a complex product takes a long time to program. Furthermore, as Peters notes, given the low programming frequency, the integrated option would only have a limited effect on operator's jobs. For the second case, the argument that 'decentralized programming would lead to too much downtime' is not worked out, but probably has to be attributed to the high programming frequency. The argument for centralized programming in the third case is, that the short cycle time does not allow parallel programming. This argument evidently overruled the rather strong attention for an integrative organization design (group production, elimination of hierarchical levels), and the presence of one CNC-machine which allows parallel programming. In case 4, there is a low repeat frequency, combined with a high programming frequency, i.e. much programming time is required. The small batches, and the fact that the complex and changing products urge operators to monitor the production process closely, leave little time to program, although the used programming package simplifies and fastens programming. There is no rotation between operator and programmer because of strongly diverging skill requirements. In the fifth case, machine downtime is again the obstacle for the creation of integrated functions, although given the high repeat frequency and low programming frequency little programming time would be needed.

The sixth case describes a machine-building factory, consisting of at least six departments or 'autonomous groups'. In two of these groups CNC-machines are placed. Four machines are used in group 'Small', and six in group 'Round'. In this last group, two machines are used for complex milling operations (these milling operations are three dimensional compared to two dimensional turning). Except for these two machines, the integrated option was chosen. Parallel programming is technically possible. There is a high programming frequency, due to the product range and the fact that programs are not stored, and can thus not be repeated. Reasons to choose integrated functions were:

1 the split and partially integrated options would add 'an additional counter', i.e. unnecessary communication between programmer and operator;
2 a separate programmer is expensive;
3 the programs are short and simple;
4 operator involvement and the fact that other options would lead to an erosion of craft knowledge.

The option for integrated functions seems to fit in with the integrative organization design, namely autonomous groups. Nevertheless, the milling operations are considered to be too complex for integrated functions. Furthermore, decentralized programming was more easy to implement in group 'Round' than in group 'Small', as 'Round' has larger batches, longer cycle times, and simpler programs, and because the machines in 'Round' change tools automatically, which leaves more time to program. In the future, the more complex programs in 'Small' may be made in a separate programming location in order to prevent downtime.

Peters' data as presented above allow to make a tentative analysis about the relationship between the output characteristics and the chosen options with respect to the division of labor. Peters himself analyzed his cases from the point of view of possibilities to improve operator's job content, focusing on the creation of integrated functions. For cases 1 to 5, Peters discusses the use of a separately located programming station, where operators can write programs. As this programming station can be used by rotation, downtime can be prevented. With respect to this solution, he makes the important observation, that in case of a low programming frequency, the inclusion of programming into the operator's function has little meaning, as this task is hardly ever conducted (1986, p. 71).

Evidently, the prevention of downtime is the dominant economic logic in all cases, irrespective of the options chosen. Preventing downtime and creating integrated functions means, that programming and operating have to be conducted simultaneously, which can be achieved by means of parallel programming, or by Peters' recommendation, namely rotating CNC-operators between operating machines and programming on a separate (programming) computer. Although parallel programming is not possible except for one machine in case 3 and in case 6, it could have been included in the other CNC-machines. An analysis of the influence of the output characteristics on required and available programming time provides a possible explanation why this feature was not included. If programs can be written during machine operation, downtime is prevented, i.e. if the available programming time is smaller than or equal to the time required to write that program. The cases provide several indications of the influence output characteristics have on the available and required programming time. The cases where the argument is mentioned are put in parentheses.

The available programming time is determined by the time to process batches, which is in its turn determined by batch size (case 4) and the cycle time of individual products (case 3). Cycle times are partially determined by product and\or process complexity (case 6). Finally, the available time may or may not partially or wholly be absorbed by other tasks than programming such as loading and unloading (case 6) or monitoring the production process (case 4).

The required programming time is wholly or partially determined by the following factors:

1 product complexity (case 1);
2 product variability c.q. programming frequency (case 4);
3 ease of programming c.q. user friendliness, which influences programming speed. Special programming packages may speed up the programming process (case 4).

These relationships are discussed in more detail at the end of this paragraph, after some additional material, relevant to this discussion, has been presented.

The study by the Economic Institute for Medium- and small-sized enterprises EIM contains three cases and a survey on CNC-machines (Poutsma, van der Staal and Zwaard, 1986; Poutsma and Zwaard, 1987 and 1989; Poutsma, 1991).

The first case is a firm employing 22 persons. It produces machine components, some in small batches, others are unique pieces. The second firm, employing 77 people, produces precision components, which are 'very varied' and are made in small batches. Similar output characteristics holds for the third firm, which has 103 employees. However, unlike the other two firms, it is not a component supplier, but produces and markets its own compressors. The nature of product range is also different from the other firms, and can be characterized as 'variations on a theme', while the firm strives for further standardization by means of group technology. It is also unique in having a centralized organization, which contrasts with cases 1 and 2. The latter have few hierarchical levels, and broad shop-floor autonomy. Poutsma c.s. stress the relationship between the output characteristics 'product range' and 'batch size' on the one hand, and the routine nature of the work on the other hand (1986, p. 75). The decoupling of programming and operating tasks is recommended against in case of complex products and a varied production, although it was also observed, that in case 2, complex programming jobs were done on a separate programming computer. In this case, the remaining 90 per cent of programs are made at the CNC-machine. Initially, management meant to program centrally, but due to resistance of operators, who felt they were being degraded to 'button pushers', and the negative consequences of this on product quality, it was decided to opt for integrated functions, especially as the complicated programs require frequent adjustments. In cases 1 and 2 there are integrated functions, in contrast to the more fractionized case 3. In the past, the latter firm had recruited many unskilled workers, whereas in the first two organizations skilled (LTS) craftsmen learned to program. In all cases, the introduction of CNC-machines led to job displacement, although nobody was fired. Other common elements are the relatively long period to make the new machine fully operational, and the stress put on a maximum utilization of machine capacity, which is uttered in attempts to increase the number of productive machining hours.

With respect to the labor market, it is reported that CNC-operators, which were all internally selected and schooled, may take advantage of their supposedly strong position. This is felt to increase an organization's vulnerability.

The EIM-survey covered 780 organizations of varying sizes. It was first conducted in 1985, and repeated in 1987. Besides CNC-machines, the computer-controlled machines in the EIM-survey include industrial robots and equipment used in process industries. It focused on qualitative changes in operators' job content. Table 4.2 contains data about the relationship between organizational size and the occurrence of integrated functions.

Table 4.2
Organizational size and integrated functions in Dutch manufacturing industries in 1985 (percentages)

	Organizational size		
Person who programs	< 20	20-99	>= 100
Operator (integrated function)	67	56	38
Other person	33	44	62
Programming location			
Machine	82	52	35
On shop-floor close to the machine	14	25	25
Elsewhere	4	23	40

Source: Poutsma, 1991, pp 44-45

Unfortunately, the data were measured at the organization rather than the machine-level. Thus, if one integrated function occurs in an organization using more than one machine, this option is scored as an integrated function. Measurements at the machine-level would result in lower percentages of integrated functions, as large organizations tend to have more CNC-machines. The exact impact can not be calculated.

The data show that the occurrence of integrated options, and programming at the machine, decreases with increasing organizational size. Even from a pure combinatorial point of view, large organizations have more possibilities to combine tasks into functions. From an organizational point of view, economies of scale can be achieved by large organizations, making a more extreme division of labor and the concentration of the programming work for more than one machine in the hands of one programmer more likely. Furthermore, the data show that shopfloor programming is not necessarily equal to the choice for integrated functions, although a close relationship between both variables is generally assumed. As shown in Table 4.3 integrated functions occur less with increasing batch sizes.

Table 4.3
Batch size and integrated functions in Dutch manufacturing industries in 1985 (percentages)

Person who programs	Batch size		
	< 100	100-1000	>= 1000
Operator (integrated function)	74	50	43
Other person	26	50	57

Source: Poutsma, 1991, p 45

Comparing briefly the results of the 1985 and the 1987 surveys, Poutsma (1991, p. 45) concludes that the percentage of integrated functions has decreased drastically (by more than 50 per cent). However, the 1987 data were gathered in a slightly different way than the 1985 data. The latter concern computer-controlled machines in general, whereas the first concern CNC-machines. In 1987, programming was done by 29 per cent of the CNC-operators included in the survey, and optimizing\correcting by 60 per cent. Assuming that all operators who program, also optimize, the percentages of integrated, partially-integrated and split options were respectively 29, 31, and 40 per cent in 1987.

The TOWES-data are a useful supplement to the EIM-data (Alders, Christis, and Bilderbeek (1988). This project was commissioned by the Ministry of Social Affairs and Employment. Both process and machine-building industries were included in the survey. As this chapter focuses on discrete manufacturing, process industries are excluded in the following. The main source is an article by Alders and Christis (1988a), which concentrates on the machine-building branch. Their classification of options with respect to the division of programming-related tasks was introduced at the beginning of this chapter. The reason to distinguish between programming on the one hand, and optimizing on the other hand, is, that programming is a creative task, requiring insight into the various aspects of the machining process, and their interrelationships, whereas testing\optimizing only require, that one understands the logic of a program, and that faults are detected. Programming is thus a far more complex task than optimizing.

The EIM-survey differs from the TOWES-survey in at least two aspects. In the first, the EIM-study took more industrial branches, and consequently more kinds of programmable manufacturing machines into account than the TOWES-study, which concentrated on the machine-building and process industries. For the scope of this investigation, the separate presentation of the results for the machine-building industry by Alders and Christis (1988a) has the advantage of the concentration on CNC-machines, while the EIM-study also covers other branches, - some relevant, others irrelevant for purpose of studying CNC-machines. Unfortunately, the EIM-data are only partially detailed for industrial branches, which makes its value for this purpose hard to assess.

The TOWES data provide more information about 246 CNC-machines in the machine-building industry, based on a survey among 180 firms, generally employing at least 20 persons. The selected firms do not make up a random sample, but were known to use CNC-machines. As more precise data are lacking, their findings may not be generalized. Possible biases include a more positive attitude toward innovations, and larger, financially stronger firms (cfr. section 3.2). By the end of 1987, the Dutch machine-building industry consisted of 1232 firms, employing an average of 64 persons (CBS, 1988, p. 11), so about 15 per cent of the branch is covered by the survey. Where felt necessary because of the small numbers of organizations covered by the survey, significance tests (Chi-square on cross-tables) are added to the original TOWES-data. Table 4.4 lists the TOWES findings with respect to the options chosen.

Table 4.4
The division of labor around CNC-machines in the Dutch machine-building industry, 1987/8 (percentages)

Option	**Firm level** (n=180)	**Machine level** (n=246)
Integrated	46	49
Partially integrated	10	19
Split	16	24
Mixed	19	
Other	9	8

Source: Alders and Christis, 1988a, p 76

The option 'mixed' indicates the choice for different options within one organization, which is not uncommon in the machine-building industry (cfr. Table 4.1). This fact, plus the divergence between firm level data and machine level data, points to the importance of this distinction. As the percentages of the integrated option differ only three per cent between both levels, in contrast with nine respectively ten per cent for the partially integrated and split options, one can conclude, that the integrated option is more coherent within organizations.

Table 4.5 presents the TOWES-findings with respect to the influence of organizational size. Although Alders and Christis draw the conclusion from these data, that an increasing firm size is linked with an increasing division of labor, the high percentages for 'mixed options', especially in the larger size categories, blur this relationship.

Table 4.5
Organizational size and integrated functions in Dutch machine building in 1987/8 (percentages); n=180

Option	Organizational size (size of workforce)			
	< 50	50-99	100-199	> 200
Integrated	67	39	34	15
Partially integrated	4	13	16	8
Split	10	17	21	23
Mixed	16	27	22	54
Other	3	4	9	0

Source: Estimated from diagram 1, Alders and Christis, 1988a, p 76

With respect to batch size and the division of labor, Alders and Christis claim the existence of a 'U-form': the highest percentages of integrated options are found at both extreme batch sizes (see Table 4.6).

Table 4.6
Batch size and integrated functions in Dutch machine building in 1987/8 (percentages); n=180

Option	Batch size				
	< 10	10-49	50-99	100-999	> 1000
Integrated	53	35	45	42	60
Partially integrated	6	15	5	16	0
Split	19	14	12	23	0
Mixed	19	31	32	18	0
Other	3	5	6	1	40

Source: Estimated from diagram 2, Alders and Christis, 1988a, p 77

From Table 4.6, Alders and Christis conclude that batch size has an ambiguous effect on the division of labor. According to them, it may have a favorable impact on integration as well as fragmentation. With small batches no coordination losses due to separate functions are incurred by integrating tasks, but downtime may occur. In contrast to this, large batches incur only little coordination losses, but leave ample time to program. However, alternative interpretations of these data may point to the existence of a U-form for the split option, leading to the exact opposite of Alders and Christis' interpretation, or a 'W-form' for the integrated option. The average number of firms per batch size category is 36, which is a small numbers situation and makes it hard to draw hard conclusions. Machine level data might have provided more clarity.

Product complexity was combined with machine complexity into a variable 'machining complexity', and operationalized into a dichotomous measure, namely two- and three-dimensional programming, the latter obviously being more complex. The relationship between this measure and the options with respect to the labor organization was not significant, as can be concluded from their data (Chi-square of 1.87). Next, different operations were related to the organizational options. More integrated functions were found with the simple operations turning and milling than with the more complex operations cuttering and machining in general. Alders and Christis give two alternative explanations for this finding. Firstly, complex programs need more programming time, and secondly, firms may judge operators incapable to program independently. The researchers are inclined to attach more value to the latter interpretation.

Programming comfort, operationalized as the level of the programming language, and the presence of automatic changing tool and workpiece, which allow more time to program, were found to coincide with an increasing rather than a decreasing division of labor. In other words: in circumstances which are favorable for the creation of integrated functions, less integrated functions occur.

They argue that there is little difference of opinion concerning the 'technical efficiency' of integrated functions, as machine-operators are better acquainted with local circumstances than programmers, and coordination losses do not occur, which is a concrete manifestation of a classical disadvantage of fractionizing labor (cfr. section 2.1).

Furthermore, a relationship between the number of CNC-machines and the division of labor was observed. According to Alders and Christis, an increasing number of CNC-machines makes the concentration of all programming jobs into one central programming function likely because economies of scale can be achieved (1988, p. 80).

In the light of MST (see Kuipers and van Amelsvoort, 1990, pp. 65-120), the findings as shown in Table 4.7 with respect to lay-out and division of labor are highly interesting.

Alders and Christis note that the differences between both lay-outs are not as big as could have been expected from the (JB: sociotechnical) literature (1988a, p. 81). Therefore, they 'are inclined to conclude [...] that the majority of firms prefers divided organizational forms if possible (depending on the number of CNC-machines and batch size) [...] the favorable conditions for task integration offered by group technology are not fully used' (1988a, 81), which can not be explained by the TOWES-study according to the authors.

Table 4.7
The division of labor and lay-out (percentages); n=246

Option	**Lay-out** **Functional (78 %)**	**Group (22 %)**
Integrated	44	49
Partially integrated	22	29
Split	34	22

Source: Alders and Christis, 1988a, p 81

As the relationship in Table 4.7 is not statistically significant (Chi-square of 5.56), the hypothesis that a group lay-out coincides with more integrated functions than a functional lay-out, has to be rejected. Possible explanations for this finding are that a group lay-out of machinery is not necessarily combined with team working c.q. (semi-)autonomous groups as prescribed in MST, or in case of such a combination, that an intra-group division of labor was established.

COB\SER-study

In 1984, COB\SER commissioned a research project on CNC-machines, conducted by the Tilburg based Institute for Social Research and the

Leiden based consultancy agency Intervisie (see: ten Have and Fruytier, 1987; COB\SER, 1988; Fruytier, 1988; ten Have, 1988; Fruytier and the Have, 1990). The effects of the organization structure around CNC-machines on the four criteria 'productivity', 'product quality', 'flexibility', and 'quality of working life' formed the central theme of the COB\SER-study. This study is to an important degree influenced by MST, and was called 'more experimental' by Poutsma (1991, p. 48). Eight cases were described, of which Table 4.8 gives some key characteristics. These data differ slightly from the original data presentation with respect to the following items:

- the use of the term 'quasi-group', which is used in the report, but not in its summarizing scheme;
- the summing up of the number of indirect tasks instead of specifying them;
- the direct tasks are condensed into the options which are used throughout the chapter;
- 'product range' is derived from 'programming frequency';
- the words 'Tayloristic' c.q. 'T-type' on the one hand, and 'New Production Concept' c.q. 'N-type' on the other hand, are replaced by an 'F' for 'fractionized' and an 'I' or 'integrative'. The T-type and N-type are characterized by, respectively, a low and high quality of working life (1988, p. 25).

The researchers experienced substantial difficulties in categorizing the eight cases with respect to lay-out and labor organization. Ship 1 and Crane were both categorized as having a group lay-out. As is explained in note 6 of their report (1988, p. 35), these cases were difficult to categorize. Crane employs about 60 people, and its factory shows 'little structure', but was labelled 'quasi-group' because of the small degree of separation between the functional departments. The machining centre in Ship 1 was placed in a group, next to three parallel lines (Ship 2 deals with one of these lines).

Similar difficulties are reported with respect to the organizational options. Here, three criteria are used for this categorization:

1 the option chosen to divide tasks directly related to the CNC-machi nes;
2 the in- or exclusion of indirect tasks;
3 the assignment of operators to fixed machines.

The last criterion is important, as rotation of operators over several machines provides learning opportunities, and leads to a broad deployabil-

Table 4.8
Case characteristics in the COB\SER-study

Case	Transport 2	Ship 1	Crane	Supplier	Barn	Transport 1	Ship 2	Diesel
Lay-out	group	quasi-group	quasi-group	functional	functional	line	line	line
Integrated\Fract.	Integ.	Integ.	Integ.	Integ.	Frac.	Frac.	Frac.	Frac.
Option	PI	I	I-PI	PI	PI	PI	PI	S
Setting tools	setter	operator	operator	operator	setter	setter	setter	setter
Indirect	7	2	4	0	0	1	0	3
Fixed assignment	no	yes	yes	no	yes	both	yes	yes
Prog. frequency	low	high	high	high	high	low	low	low
Prog. complexity	high	low	low	high	low	high	low	high
Batches	10-100	2-30	100-300	1-5	20	3000	≤ 100	10-50
Parallel program.	no	yes	no	yes	no	yes	no	yes

Source: COB\SER, 1988, pp 54-55 and p 104

ity of these operators. It was decisive for the classification of Supplier and Barn, both having a functional lay-out, as, respectively, integrative and fractionized. However, except for Supplier and Transport 2, all cases, including the other two integrative organizations, have fixed assignments. All organizations with a group lay-out are characterized by a relatively broad functions, but these are not necessarily integrated. All partially-integrated functions are characterized by a cooperation between operator and programmer with regard to program testing.

It is striking that the main difference between integrative and fractionized organizations in this study concerns the indirect (in the sense of not directly machine-related) rather than the direct tasks. In general, the integrative organizations are characterized by a larger number of indirect tasks (although Diesel forms an exception in this respect), and by the fact that operators tend to set machines rather than a specialized machine-setter (with the exception of Transport 2). Fruytier and ten Have stress that '[t]he quality of work in a production structure with flexible production groups can remain high because 'organizational' tasks are added to the tasks directly related to the machine, with adequate control capacity for the employees in the groups' (1990, p. 257).

The categorization problems confirm the polar instead of dichotomous character of the distinction between integrative and fractionized organizations.

With respect to the absence of a statistically significant relationship between organizational option and lay-out in the TOWES-project, the two cases in Transport are interesting. Both use group work, and little differences exist with respect to the machine-related tasks. As a whole, in Transport 2 the division of labor is less extensive than in Transport 1, mainly because of the assignment of indirect tasks to the operators. The example shows that group work can also be combined with a line lay-out. These choices can be attributed to Transport's project 'quality of working life'.

Several observations can be made concerning the output characteristics, and their influence on organizational options (with respect to direct tasks). However, with six out of eight cases having partially-integrated functions, the relationships are tentative. The four cases with a high programming frequency use the I- or PI-option, while the remaining four cases have partially-integrated or split functions. This relationship is not very strong, as can be concluded from the dominance of the PI-option irrespective of the programming frequency. A similar relationship can be found between programming complexity and organizational option. Ship 1 and Crane, the two organizations where operators program, have a low programming complexity. In case of Crane, which also employs a separate programmer, operators take charge of the simpler programs. Batch size, which is remarkable low on average, seems to be irrelevant to the organizational option.

The possibility of parallel programming is not used in three out of four cases, whereas in Crane, the only organization having at least partially chosen for an integrated function, no parallel programming is possible. This organization probably uses a separate programming location to allow operators to program.

One may hypothesize that differences concerning the division of labor occur when more than one shift is used. Two publications by Dirne (1990 and 1991) in the field of (quantitative) operations research, provide an indication in this direction. The Philips Machine Factory in Almelo purchased two machining centres in the mid-1980s, which for efficiency reasons were used in three shifts, the third shift being unmanned. The production capacity of the unmanned and manned shifts had to be split with respect to the output produced. In the first place, products run during the unmanned shifts have to be less complex than products run in the manned shifts. During unmanned production, disturbances in the production process can not be resolved. Simpler products that were first produced flawlessly during manned shifts, reduce the chance that disturbances occur considerably. The more complex products remain for the manned shifts. In the second place, long batches have to be produced during the unmanned shift. As the running time of these batches is always shorter than eight hours, i.e. until the first manned shift starts, the number of unmanned hours is maximized by planning batches, that require a long processing time.

Labor market

Before proceeding to a discussion of issues at the labor market for CNC-operators, a brief discussion of the relevant parts of the Dutch system of technical education and vocational training is given (CEDEFOP, 1984; Dankbaar, 1990; Mason, Prais and van Ark, 1992). The lower level of technical education is the LTS-level, 'Lower Technical Schooling'. The LTS takes four years after the primary education, which normally ends at the age of 12 years. At the LTS, students can choose between grades, namely A, B, C and D (rising degree of difficulty). After graduation, students with the higher grades (C and D) often continue their education at the intermediate-level of vocational training MTS. Alternatively, they may take up a position as an apprentice in the so-called 'leerlingwezen'. An apprenticeship usually lasts two years. Upon successful completion a certificate is awarded. Apprentices may work anywhere between a half day and four days per week in a firm; the remaining days are used for theoretical and practical schooling. Apprentices are either employed by firms or by regional schooling centres. About 20 per cent of the firms in the metallurgic industries take part in educating employees by means of the national apprenticeship system (van Zwol and Bots, 1992, p. 64).

The LTS is oriented at schooling craftsmen, whereas MTS schooling has more theoretical components. MAVO-graduates (MAVO = intermediate general secondary schooling) are also entitled to enter the MTS. As far as CNC-machines are concerned, an important distinction is, that LTS-graduates often have acquired practical machining experience, whereas the knowledge of MTS-graduates is more of a formal and abstract nature, for instance programming. Typically, MTS lack practical machining experience, and LTS lack programming knowledge. If no integrated options exist, communication between programmer and operator is necessary as the first person has to formalize the latter's experience.

The TOWES-project paid extensive attention to the labor market. Table 4.9 shows the required education related to type of machine, and division of direct labor. As shown in this table, educational requirements rise after the introduction of CNC-machines, which is related to the organizational option chosen. For integrated functions, 67 per cent of firms judge MTS as necessary, which percentage drops from 45 for the partially integrated to 25 for the split option. Programmers are required to have MTS in 98 per cent of all cases. As could be expected, there is a statistically significant relationship between the option and required education (Chi-square of 31.06, which is significant at an uncertainty level of .001). Nevertheless, 25 per cent of firms judge MTS as necessary even in case of the split option, although the absolute number of cases is rather small (17 or 18, depending on rounding).

Table 4.9
Required education of conventional and CNC-operators in Dutch machine-building industries, 1987/8 (percentages)

Required education / Machine	LTS	LTS+	MTS	Other
Conventional machine		90	9	1
CNC-machine	17	36	45	2
Option				
Integrated		32	67	1
Partially-integrated		54	45	1
Split		75	25	0

Source: Alders, Christis and Bilderbeek, 1988, p 20 and p 24

'LTS+' refers to LTS-graduates who took additional courses, generally as apprentices.

The wide-spread tendency to ask for an MTS-level led to a high demand on this segment of the labor market, which could not be met by the supply of MTS-graduates. Consequently, they occupy a strong position at the labor market, which enables them to select attractive CNC-jobs, for instance programmer or an integrated function, leaving less attractive tasks for lower skilled persons. This development creates a ceiling in the internal career trajectory for LTS-graduates, who are no longer able to qualify internally and to gradually learn programming. An additional problem was formed by the at that time existing lack of suitable courses which allow LTS-graduates to learn to program. Based on the tendency to create attractive jobs for MTS-graduates, Alders and Christis expected an increasing division of labor around the CNC-machines in the future, especially if economies of scale can be achieved when more CNC-machines are purchased (1988a, p. 80). This concentration of programming makes the firm vulnerable, as the CNC-programmers can easily switch jobs.

As a part of the COB\SER project, ten Have (1988) pays attention to the recruitment of CNC-operators. He concludes, that operator's factual skill level, nor the desired skill level by management are related to job content. Operators possessing skill levels ranging from unskilled to MTS perform the same kind of work. In at least five cases, management would prefer to have higher skilled operators than are present. The fulfilment of these wishes would lead to a further underutilization.

The firms' recruitment policies offer four explanations for these phenomena:

1 uncertainty effect;
2 start-up effect;
3 anticipation effect;
4 image effect.

The uncertainty effect stems from the fact that the introduction of CNC-machines involves many uncertain elements. To ensure that these uncertain elements are handled well, firms tend to raise skill requirements. In case of a start-up effect the new processes have to be learned during the first phases of using CNC-machines. The effect is temporary as these processes gradually become routine. The anticipation effect resembles the uncertainty effect in that uncertainty about future requirements tend to raise the demand for skills, but this uncertainty has not yet materialized. According to the image effect firms derive status from a high skilled work force.

These combined effects result in a tendency to ask higher skilled operators than necessary. In the longer term, ten Have expected a self-correcting process: the labor market scarcity would ultimately necessitate organizations to lower skill requirements. If the FME-data presented in section 2.4 can be generalized, it can be concluded that this process had not started by 1990.

Conclusions and discussion

The percentages of integrated functions found in both surveys vary considerably, namely 29 per cent in 1987 according to the EIM-survey and 49 per cent in 1987/8 according to the TOWES-study. As their data partially concern different populations and different machinery, use different operationalizations, and the generalizability of both samples is questionable, the comparability of these data is doubtful. The differences must at least partially be attributed to the factors mentioned instead of to differences between manufacturing industries and\or periods. Nevertheless, both studies proof that a large variety of organizational options exists around CNC-machines, and that some of these options are significantly correlated with certain characteristics. Less integrated functions are found when:

- the organizational size increases (EIM, probably TOWES);
- the number of CNC-machines increases (TOWES);
- two-dimensional operations are compared with three-dimensional operations (TOWES);
- the programming comfort increases (TOWES).

Whereas the first three relationships could have been expected, the opposite relationship could have been expected in the last case, because more user-friendly ways of programming lower skill requirements *ceteris paribus*, lowering a barrier for the formation of integrated functions (cfr. section 2.2). In the words of the TOWES-researchers: 'favorable conditions for the creation of integrated functions exist' (Alders and Christis, 1988, p. 83). Probably, the *ceteris* are not *paribus*: more factors than just user-friendliness change simultaneously, or in statistical terms, multicollinearity occurs. In this case, user-friendly programming languages and packages which increase programming comfort, which are probably more expensive than less user-friendly ways of programming, may be purchased for more complex operations, i.e. when the need for user-friendly ways of programming is higher (see case 4 of Peters). Furthermore, expensive (in terms of machines as well as programming method) systems may preferably be entrusted to high skilled employees. A similar problem occurs with respect to the connection between organizational size and the number of CNC-machines.

For some other characteristics, the evidence was either not significant (machining complexity, lay-out) or inconclusive (batch size). In the TOWES-study machining complexity was a composite measure, which makes it impossible to draw a clear conclusion. The problem of interrelated aspects may have occurred at least at one other point, namely with respect to organizational size, batch size, and the occurrence of integrated options. Organizational size was demonstrated to be negatively related to the occurrence of integrated functions, which relationship was also hypothesized for batches size. If smaller organizations produce on average smaller batches, as is often assumed (Stokman, de Groot and Leurs, 1989), it remains inconclusive to what factor the smaller number of integrated functions has to be attributed. If the number of machines, which is probably positively related to the organizational size, is also taken into account, the total picture gets more complicated. Organizational size was shown to be positively related to batch size, which makes a correlation between batch size and the number of machines likely.

Concluding, the surveys provide some information concerning some relationships between certain characteristics and the number of integrated functions, but the exact patterns of causality can hardly ever be established. An additional problem is formed by the limited comparability of both surveys. To gain more insight in the possible underlying relationships, data from the case-studies are discussed next.

Perhaps the most relevant question with respect to the assessment of developments in the work content of CNC-operators is whether only machine-related tasks have to be taken into consideration, as in the TOWES-study, or indirect tasks are important as well, as in the COB\SER-study. In the latter study, the concept of jobs with a high decision latitude, as well as job demands, is important, which is in the MST-tradition. The conventional craftsman is a stereotypical example of such high quality jobs. Such a craftsman is often assumed to have been responsible for almost the total production process, i.e. from reading the technical drawings to the actual manufacturing and quality control. With the introduction of numerical control, the craftsman's machining knowledge was formalized in programs, which were first recorded on punch tapes (NC) and could later be stored in computers and on floppies (CNC). The transition from conventional machines to computer-controlled machines meant the formalization of machining knowledge. The new technical system thus led to a change in the organizational task pool. The machining process stays the same, but the form of control, i.e. the way of operating the machine, changes. Programming and optimizing are added to the organizational task pool, and the former way of control, i.e. manual intervention by the machinist, is replaced. This formalization of machining knowledge requires other skills: the machinist's machining experience is replaced by the knowledge required to write programs. In the case of CNC-machines, this generally concerns the ability to think in an abstract way in order to formulate mathematical statements which make up a program. The introduction of NC- and CNC-programs made it possible to separate the conceptual tasks, roughly the trajectory from reading drawings to writing programs, from the executing tasks, i.e. all tasks related to machine operation. Some authors seem to assume that this option of separating both kinds of tasks is always chosen. For instance, Adler and Borys assert that 'With NC the combination of machine and machinist is replaced by a machine, a computer, a human operator, and a programmer' (1989, p. 377). Similar statements can be found in Noble

(1978) and Wilson and Buchanan (1988, p. 366). However, such a separation of tasks is not an automatism, but the result of applying consciously or unconsciously organizational design principles, i.e. of dividing the tasks of the changed organizational task pool via design rules over employees. Empirical results as discussed above for the Netherlands and as found abroad (Dodgson, 1987; Keefe, 1991; Kelley, 1991; Nielsen, 1992; Sorge, 1985), as well as discussions about the boons and merits of decentralization and centralization of programming c.q. integrated functions (van Amelsvoort, 1985; de Korte, 1991), prove that a large variety of options is possible and actually chosen, and is also subject of conscious decision-making processes. Even if one particular option is chosen in the majority of cases, as in Shaiken, Herzenberg and Kuhn (1986), it remains one dominant choice out of a range of possible choices.

With respect to the initial question, the form of control of the machining process has changed from manual control to programming with the replacement of conventional machines by NC-machines. The locus of control of the machining process depends on the person who is responsible for, respectively, manual control and programming. With conventional machines, this person is the machine-operator. The introduction of CNC-machines has introduced a new option, namely the assignment of programming to either the machine-operator or a specialist c.q. programmer. Control of the machining process is a vital part of a craftsman's job, who works autonomously. However, for this purpose the craftsman needs to be in command of more than just programming: he must also be allowed to take decisions concerning activities such as work distribution, sequencing of orders, and choice of tools. In the COB\SER-study this position was adopted for CNC-operators, but it also showed that the inclusion of indirect tasks in their jobs does not necessarily coincide with integrated jobs. It is doubtful whether such functions can be considered as equal to those of craftsmen. CNC-craftsmen should have an integrated function to control the machining process itself as well as indirect, regulating tasks concerning the organization around the machining process. Yet comparing conventional craftsmen with CNC-craftsmen, one has to realize, that those indirect tasks must also have been part of the jobs of conventional craftsmen. In other words: there are also different organizational options around conventional machines, especially with respect to the division of tasks, which are not directly related to machines such as work distribution. Specialized tool setters were common with conventional machines. It has to be kept in mind, that the period in which conventional machines were being used was characterized by fractionized designs (see section 2.1). The fact that different organizational options also exist with conventional machines has not always been realized, especially not if the romanticized picture of the craftsman is kept in mind. An adequate comparison of the situation before and after the replacement of conventional machines by CNC-machines is not possible, when different tasks structures are compared. What changes is the form of control of the machining process, and the introduction of the possibility to change the locus of control of this machining process. Whether or not the locus of control is assigned to the CNC-operator, and whether or not other tasks are included in his function, is a question of exercising organizational options.

This standpoint necessitates a slightly more nuanced view than formulated by hypothesis 1, which states: 'the application of organizational design orientations is independent from the technical system(-s) used'. The new form of control (C)NC not only creates a new organizational option, but it also sets conditions within which the organizational design choice is made. The transition from conventional to numerical control has implications for the skill levels used, as stated above. Abstract, mathematical thinking is generally necessary to write programs. If conventional machinists do not possess these skills, the programming is performed by somebody else, i.e. a specialist. The development of 'human centred' programming packages (Brödner, 1991) which use graphical techniques to program and fit the ways of thinking of skilled machinists, is important in lowering this barrier. However, the use of such user-friendly packages is by no means a guarantee for the formation of integrated functions, as was shown in the TOWES-study, where such programming comfort was negatively related to the occurrence of integrated functions. It needs to be borne in mind, that user-friendly programming is advantageous to fractionized organizations as well. Therefore, Brödner emphasizes an integral approach to human centred systems, of which the development of technical systems forms a part (also see section 2.2). Parametric programming, i.e. filling in the value of parameters into pre-programmed macro's, is fairly similar to human-centred programming in this respect. It lowers skill requirements posed on the CNC-operator, who is still in control of the machining process, but only for a limited range of products, namely those products for which macro's have been written. Whether or not this can be considered as an integrated function is subject to discussion, but the example shows that even this seemingly simple question is more complex than it appears to be at first sight.

Finally, as de Korte (1991) stated, parallel programming allows programming during machine operation. Again, using this option is optional. The cases in the COB\SER-study and the study by Peters show that integrated functions may occur even when parallel programming is not possible, whereas parallel programming is not necessarily combined with integrated functions. All combinations seem to be possible, which probably depends to an important extent on output characteristics (see below). Unfortunately, no data are available on the diffusion of programming methods in general, and parallel programming specifically.

From an MST-point of view, the absence of a relationship between lay-out and organizational options in the TOWES-study is interesting. The COB\SER-cases indicate a positive relationship between the number of indirect tasks and the group lay-out, but the case 'Diesel' combined a line organization with the split option and a relatively large number of indirect tasks. This lends support to Fruytier and ten Have's position (1990), that indirect tasks are important to judge the work content, but it also points to a much looser connection between lay-out and the division of labor than is assumed, or rather prescribed, by MST. The bundling of machines in a group lay-out is sometimes connected to an integrative design orientation, but it is by no means a necessary precondition for the creation of team work.

Striking similarities in the case-studies are the dominance of the partially-integrated option compared to the surveys, and that optimizing/correcting is often jointly done by a specialist programmer and a machine-operator. Probably, this last fact can be explained by a practice, that the machine-operator signals necessary adaptations while running a program, which are then implemented into the program by the programmer. The dominance of partially-integrated function may be a statistical artefact, but may also point to a larger use of integrated functions in the machine-building industry, which was the subject of the TOWES-study. Where integrated functions occur in the cases, it concerns simple products. The programs for more complex products were written by specialists (with the exception of the first EIM-case). With this statement, the discussion of the role of the output characteristics has started.

Hypothesis 2 states: 'The lower product complexity, the smaller product variability, the smaller the product range, and the larger the batch sizes, the less likely it is, that integrated jobs are found'. The role of product complexity has already been discussed partially. Whereas Poutsma recommended the use of integrated functions in case of complex products, the cases almost consistently show that less integrated functions occur with an increasing product complexity, despite the disadvantage of necessary intensive coordinating activities between operator and programmer in case of partially-integrated and split functions. An increasing product complexity has also implications for the length of the programming and processing times. The resulting picture is that there are arguments for a positive as well as a negative relationship between product complexity and the use of integrated functions. Unfortunately, quantitative Dutch research data concerning this relationship are not available.

The study by Peters allowed to make an inventarization of factors which are likely to influence positively and negatively the use of integrated functions. The analysis which was made in the beginning of this chapter can be supplemented by the analysis of Alders and Christis (1988a, pp. 81-83), who seem to have assumed in their study that parallel programming is always possible. These analyses start from the observation, that machine downtime has to be prevented. Although the economic logic of such a policy is contested (Kaplan, 1983) as it may lead to long delivery times, its prominence is empirically demonstrated in the cases of Peters. Downtime can be prevented if programs can be written during machine operation, i.e. when the required programming time is shorter than the processing time c.q. the time available for programming. The required programming time increases if product complexity increases, product variability and hence the programming frequency increase, and process complexity increases. The available programming time is influenced positively by long batches, complex products and hence long cycle times, and the degree of automation, and negatively by the time absorbed by other tasks than programming (an organizational factor). So far, the consequences of (mainly) output and technical characteristics for the occurrence of integrated functions have been discussed. As indicated above, some of these characteristics also influence the skill requirements and the capital intensity, which are both negatively related to the occurrence of integrated functions. Figure 4.1 gives an overview of these relationships.

Most of these relationships depicted in Figure 4.1 are derived from the empirical material discussed above, others were proposed as plausible explanations for unexplained empirical findings. In some cases, the sign, positive or negative, of the relationship may be subject to discussion. Although being tentative and simplified, Figure 4.1 makes clear, that unidirectional and direct relationships between most output characteristics and some technical characteristics on the one hand, and the occurrence of integrated functions on the other hand are oversimplifications. The relationships hold *ceteris paribus*, but the *ceteris* are hardly ever *paribus*.

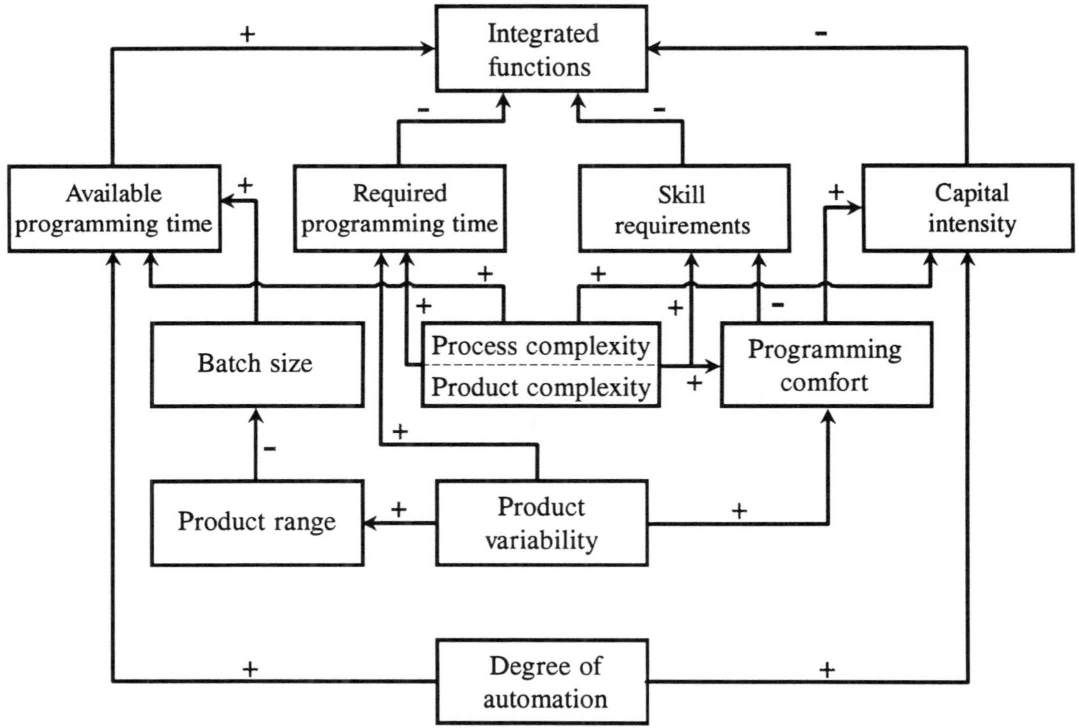

Figure 4.1: **Output and technical characteristics in relation to the occurrence of integrated functions in case of parallel programming**

Some of the mentioned characteristics such as product complexity and programming comfort have contradicting influences. Yet, Figure 4.1 does still not fully explain all findings, as organizational considerations (see above) and other contingencies are not included. For instance, the evidence with respect to batch size was inconclusive: Table 4.3 shows a negative relationship, but this was not confirmed by the data in Table 4.6 (although the negative relationship is generally found in international research as in Sorge, 1985), and Figure 4.1 even points to a positive relationship as longer batches allow more programming time. This puzzle may be resolved when organizational size is taken into account: small organizations tend to have a lower division of labor c.q. more integrated functions than large organizations. Furthermore, Figure 4.1 concentrates on the available programming time relative to the required programming time, but in an absolute sense less programming time is required if product variability decreases.

If statistical analyses show some significant correlations between the aspects mentioned in Figure 4.1, they may be interpreted as the dominance of a specific relationship rather than as proof for an unidirectional relationship, or other factors such as organizational size and the number of CNC- machines have to be taken into consideration. Hence, the second hypothesis has to be dismissed as a similar oversimplification. It may be impossible to disentangle the clew, whereby relationships as depicted in Figure 4.1 provide the ex post explanation of empirical findings. Anyway, any 'one best way' is on the far side.

With respect to hypothesis 3 concerning the use of grouping techniques to reduce the product range and its consequences for organization design, little empirical material was found with the exception of the third EIM-case. In this case, standardization by means of group technology was strived for, and product range concerned small differences in basically similar products. Although this organization was characterized as fractionized, both characteristics may be viewed as different manifestations of the same organizational policy, rather than as being directly related. Nevertheless, it is intriguing to note that the use of group technology, which is traditionally associated with integrative work organizations ('team work'), may be used to change output characteristics, namely to reduce product range and increase batches, in order to create favorable conditions for fractionized organizations. As the empirical material shows there are considerable variations with respect to the output characteristics handled by CNC-machines. The argument that CNC-machines, being 'flexible', i.e. capable of handling a large product range, are used in a wrong way when large batches are produced, has to be viewed critically, especially if techniques such as group technology are used. It is a misconception to think that CNC-machines must necessarily be used to handle a large product range. CNC-machines provide the option to be used in such a way, but the use of this option is critical. Other reasons than handling a large product, such as reproducibility and efficiency, may also be reasons to purchase CNC-equipment.

The suggestion in Dirne's work that there may be differences between shifts with respect to the output characteristics is worth looking into (Benders, 1992), but there are hardly any empirical data concerning this aspect. The matter becomes more important when capital intensity, and consequently pressures on machine utilization and the number of operating hours, increase (Mueller and Purcell, 1992). Such a policy is likely to have consequences for the division of labor.

Concerning the labor market, it was concluded in both projects that LTS-graduates are able to learn to program. This casts serious doubts on the rationality of a recruitment policy directed at MTS-graduates, even if this leads to a further division of labor. Given LTSers' (potential) ability to program, a recruitment policy aimed at MTSers has substantial disadvantages: scarce programming knowledge gets lost when the MTS-programmer leaves, MTSers are paid more than LTSers, the internal labor market is disturbed, as well as the disadvantages of more fractionized organizations, most notably coordination losses and underutilization of LTSers. Furthermore, this policy may not be sustainable, given persisting shortages at the MTS labor market segment, and given the MTSers' career perspective of entering positions in lower and middle management (cfr. FME, 1990c). As an example ten Have mentions DAF (1988, pp. 68-69). Having started an integrative design project with group work, MTSers were felt to be needed on the shopfloor. The groups were composed of MTS newcomers and sitting LTSers on a basis of equality. Soon, the MTSers started to look down on the simple, routine work which was part of their jobs, which threatened to lead to an undesired intra-group division of labor; MTSers also started to use every opportunity to switch to other jobs, either in lower management or a supportive job. These problems led DAF to a change in its recruitment policy, which was once again directed at LTSers, who were to receive additional schooling.

The question rises whether the intellectual skill level, certified by a school diploma, is the only selection criterion. Possibly, a perceived more positive labor attitude of MTSers compared to LTSers plays a role next to intellectual abilities (cfr. Gerlach, 1991). The wide-spread belief in 'flexibility' and 'changing environments', and the continuing introduction of new technical systems may have increased the influence of ten Have's uncertainty and anticipation effects, especially when the ability to quickly learn to function in a changed production environment gains importance. Because of this, his expectation about the self-correcting process on the labor market may not have come true, and employers will continue to ask over-qualified personnel. The FME-data presented in section 2.4 point in this direction. The evidence indicates a confirmation of hypothesis 4.

In conclusion one can agree with Alders and Christis (1988a, p. 80), who expected the division of labor around CNC-machines to increase rather than decrease, although some other factors may have counteracted this development. In the meantime, the lack of appropriate programming courses at the LTS-level has been resolved. Finally, the market availability of more user-friendly programming systems (more parallel programming, simulation, and parametric programming) has lowered the skill level needed to program, WOP being a fashionable example. Yet, an increasing number of CNC-machines, the use of CNC-machine for more complex products and processes, pressure on the utilization of capital intensive machinery, further integration of product design, work preparation and programming (Ewers, Becker and Fritsch, 1990) and the labor market position of programmers may be stronger factors. Whether the current campaign of the Dutch Labor Inspectorate to promote the creation of integrated functions will be successful, remains to be seen in the light of the multitude and interrelatedness of factors that play a role in these decisions.

4.2 Industrial robots

The empirical literature on 'work design around industrial robots' in the Netherlands is often quite circumstantial or outdated. A 1982 survey by Laurentius, Timmerman and Vermeulen (1982), one case in the EIM-study discussed in the previous section, an article by Huijgen and Pot (1987) containing a concise description of Volvo's (now NedCar) spot welding line, occasional references in Vrakking (1988), a study by a vocational training institute (Haring and de Mare, 1991), and a paper by Mossink and Peeters (1992) make up the harvest.

The survey by Laurentius, Timmerman and Vermeulen (1982) contains some information about the early history of Dutch robotics, at the time when, according to supplier data 71 robots were used by 46 firms. Of these, the survey included 58 robots in 38 firms, mainly in the metallurgic industries. Over half of the firms employed between 100 and 1000 persons. The major applications were arc welding (21 robots), spray painting (16 robots) and handling (13 robots). Generally, the robots replaced manual operations, which was seen as a direct motive in 51 cases. The average labor saving was 1,25 person per shift per robot. The number of operating hours was hardly ever extended. Of the 38 firms, 26 operated only day-shifts, 7 used a two-shift system, and the three remaining firms worked three shifts. The four most important reasons were; improvement of the working environment, productivity increases, (product) quality improvement, and a lack of qualified personnel. The researchers conclude, that robots are generally used to replace unattractive jobs, for which it is difficult to find personnel (1982, 9).

With respect to output characteristics, the following conclusion is drawn: 'Apparently, few programs are used (half [of the robots] uses less than 10 programs) during relatively long periods (generally changes occur after several days) with fairly short cycle times (for half [of the robots] less than two minutes)' (Laurentius c.s., 1982, 54). Furthermore, 'flexibility' was listed ninth as a purchasing motive. Small and medium-sized batches were most common (63 per cent), although batch size was not quantified. Table 4.10 provides details concerning the division of labor.

Table 4.10
The division of labor around Dutch robots in 1982

	Programming	Operating	Maintenance
Robot-operator	21	32	5
Craftsman	13	8	0
Foreman and higher	17	3	2
Maintenance technician	2	0	46
Supplier	0	0	7

Source: Adapted from Laurentius c.s., 1982, p 61

Unfortunately, these data can be interpreted in more than one way, as no information is given concerning the combination of these tasks into functions, and the omission of optimizing. However, it can be noted that 11 robots are not operated by workers with the title 'robot-operator', but by somebody else (craftsman, foreman) who probably occupies a higher hierarchical position and\or status. Furthermore, 15 robots do not seem to be operated at all as they are not included in Table 4.10, which probably includes all the handling robots. Finally, given the output characteristics presented above, one may assume that the average programming frequency at the time was low, in which case the inclusion of programming into an operator's job adds little to job content (cfr. Peters, 1986). However, drawing more exact conclusions based on these data seems speculative.

Huijgen and Pot (1987, pp. 301-303) give a brief description of Volvo's spot welding line, which was at the time of the investigation just being implemented. In this department, pre-formed steel components are joined to a limited number of car bodies by means of spot welding.

The product in process as well as many components are automatically transported between work stations c.q. the spot welding robots. When fully operational, the robotized line was to replace the manual welding stations, which was to lead to considerable job displacement. However, the effects on the labor organization and job contents were not clear at this time. Although the labor costs of the 'average production worker' were taken into account by the cost-benefit analysis, social and organizational criteria did not play a demonstrable role in the design process. Global notions about the nature of needed (human) functions sufficed. Only when a significant part of the new equipment was being installed and tuned, the hierarchical official concerned started to work out a preliminary potential function structure, which formed the basis for several months of discussions in which representatives of the personnel and organization department, and line management took part. Later, selection and recruitment issues were discussed.

A year later, a temporary organization structure was decided upon, which was to be replaced by a permanent structure in one or two years' time when the line would be fully operational. The permanent structure was still subject to discussion, without even a clear view of the major function, namely the robot operator. The temporary function structure was as follows:

- robot operators who each took care of three spot welding robots.

Their tasks included optimizing[1], quality control and simple maintenance;
- loaders, who were to supply and feed components into the line;
- manual welders for work which was not yet automated as well as to correct faults;
- group leaders;
- maintenance technicians for specialized maintenance jobs;
- mechanics\programmers to program.

Except for the robot operators, who were required to have an MTS-level, all personnel was internally recruited. The maintenance technicians were given internal additional schooling, which (obviously) was not necessary for the loaders and manual welders.

Huijgen and Pot point to the integrative approach that was used to create the function of robot operator. They also conclude that a dual function structure came into existence, with highly skilled robot operators and maintenance technicians on the one hand, and loaders and manual welders, performing routine jobs, on the other hand. The routine work was expected to disappear under influence of further automation.

The organization is known to have introduced task groups in the Born plant, but systematic data are not available.

The following case description is derived from the EIM-study, discussed in the previous section (Poutsma et al., 1986). The firm under study was one of the first Dutch adopters of robots. It employs about 50 people, and is located in a small village. Being family-owned, the firm is characterized by a centralized organization structure, but few has hierarchical levels. Most work is routine (loading and unloading) and requires little schooling. Because of this, workers can easily be interchanged, which feature is extensively used. Being scarce employees, the welders are generally put to work in the welding department, but they rotate between manual welding and operating the robot.

The robot is used for arc welding, more specifically to weld underframes for barrows. which is classified as a simple product with a short welding time. Over the years, the robot had to be re-programmed twice because of a changed frame design. Thus, the programming frequency is very low. Batches are 'several thousands', and cycle times are short. Most work is routine. However, the robot's programs need to be frequently adjusted c.q. optimized. Both programming and optimizing are performed by the technical director, as the welders have little schooling and programming requires abstract knowledge. The lack of experience, and the primitive programming method led to considerable start-up problems. The two fixtures were made internally.

One maintenance mechanic \ foreman was supposedly able to program, but he hardly ever actually programmed. The robot operators did not require additional schooling, but were instructed by the technical director. Labor saving and constant reproducibility were seen as important advantages. Overall, the introduction of the robot led to little organizational changes. Compared with some automatic spot welding machines placed in the same department, more tasks have to be performed around the robot. Furthermore, more components need to be placed more accurately in the fixture. The work at both kinds of equipment is routine, repetitive, and short-cyclic.

Vrakking (1988) describes briefly fourteen projects on the implementation of several forms of manufacturing automation. Three of these contain some occasional references to robots as a part of advanced manufacturing systems. ATAG's Lichtenvoorde plant uses three robots in a production line for furnaces, in Schlumberger's Dordrecht plant, four robots are used in an assembly line, and in Roermond, the Staalcentrum Roermond operates an advanced cutting system, of which a cutting robot forms the central part. The common characteristic in these cases is the fact that considerable job displacement took place. At ATAG, the workforce was reduced from 20 to 7 persons, Schlumberger's assembly system meant to incorporate 'a minimum of human activities' which resulted in a loss of 11 positions, and in Roermond, productivity was increased so drastically, that the three-shift system was replaced by one shift, and 10 out of 18 persons were replaced. At Schlumberger, labor saving and the prevention of the creation of monotonous, repetitive jobs were arguments for this job displacement.

With respect to job design, assembly work was replaced by supervising and operating at Schlumberger. In Roermond, heavy work such as carrying work pieces was eliminated, and replaced by machine supervision. Data on ATAG are lacking.

A more comprehensive, yet selective overview is given in Haring and de Mare (1991), whose aim it was to investigate whether the use of industrial robots had lead to a need for specialized courses in the national apprenticeship system. Their project covered 26 organizations using a total of 454 robots. Large organizations are overrepresented; the two largest organizations use a total of 383 robots, mainly for handling (152), spot welding (110) and assembly (112). The remaining 24 organizations had

71 robots; 36 of these were used for arc welding. Robotized welding was found in 17 organizations, of which 15 had no other process robotized. Compared with other applications, the production environment of welding robots (mainly arc welding) is more often isolated and less technically advanced. In almost all organizations, welding robots are fed manually; the percentage for other robots is 50. Table 4.11 contains data concerns the minimum and maximum batch sizes for both groups of robots.

Table 4.11
Batch sizes for two groups of industrial robots

Application	Minima \ maxima	Size categories of batches		
		< 1,000	1,000-25,000	> 25,000
Welding (n=17)	Minimum	14	1	2
	Maximum	7	7	3
Other applications (n=11)	Minimum	4	4	3
	Maximum	2	2	7

Source: Adapted from Haring and de Mare, 1991, p 10

It can be inferred from Table 4.11, that the average batch size is fairly long for all applications, although welding robots tend to produce shorter batches than other robots. Still, only 7 out of 17 welding robots are not used for batches of more than 1,000 products. It has to be noted, however, that large organizations, which are typically associated with large batches, are overrepresented in this study.

Although Haring and de Mare distinguish five different 'job profiles', only two of these, namely the operator-programmer (generally integrated functions) and the regular operator (partially-integrated and split functions), spend more than five per cent of their time on robot related work. The exact percentages range from 90-100 per cent for welding robots (both groups of personnel), 10 to 25 per cent for operator\programmers of 'other applications' and 1-100 per cent of the regular operator. Nine organizations which only use welding robots have operators\programmers (60 per cent), compared to 22 per cent for the total population. The organizations employ a total of 25 persons in work preparation. These people often program the robot.

Mossink and Peeters (1992) visited four robot-using plants on behalf of the Labor Inspectorate, which forms a part of the Ministry of Social Affairs and Employment, and is responsible for the control of labor laws. The Labor Inspectorate ordered the researcher to prepare a publication to instruct robot users about the effects and consequences of introducing robots on job content and the working environment.

Table 4.12 contains the data on 13 cases as far as they concern operator's job content (a fourteenth case B6 is left out, as it concerns a training station; furthermore, the original sequence was altered in order to group applications).

Table 4.12
Operators' jobs around 13 robotized workspots

Case(-s)	Application	Job content
A 1	Cleaning	Supervision
A 2-3	Handling	Supervision; once a day (un-)loading buffer
D 14	Palletizing	Supervision; occasional pallet removal
B 4a\b	Spot welding\ brushing	Supervision; (un-)loading; optimizing
B 5	Spot welding	As above plus quality control and after-treatments
B 8	Laser cutting	Supervision; (un-)loading
B 7	Arc welding	Supervision; (un-)loading
C 9-10a 10b-11	Arc welding	Integrated functions plus manual welding, maintenance, and parts supply
D 12-13	Arc welding	(Un-)loading

Source: Adapted from Mossink and Peeters, 1992, p 5

Cases B4a\b and C10a\b are both work cells containing two robots.

The authors described some further effects on job content. In cases C9-10-11, 'fairly complete jobs' exist, 'though programming and maintenance take little time' (as far as programming is concerned, this is an indication that few new products are introduced). The elimination of repetitive work was achieved in cases A1-2-3 and D14. In cases A2-3, the robots are used to (un-)load hot workpieces from\into furnaces. The latter cases form a contrast with cases B4, and D12-13, where monotonous jobs were created. 'Little autonomy' is a characteristic of cases B4-7-8, and D12-13. Few opportunities for contacts are problems in cases B4-5-8, and D12-13. Isolated works spots are mentioned in cases A1, B4-5-8, C9-10-11, and D12-13. In case C9 this is less problematic as long cycle times allow the operator to leave the work station for a short period.

The improvement of working conditions by robotizing the work, was the main motive to implement robots in case A, be it that recruitment problems were mentioned in connection with this. Recruitment problems also motivated firm C to buy arc welding robots; the professional opportunities offered by robots were supposed to attract welders. However, later the welders felt degraded: after having programmed the robot, repetitive (un-) loading remained. Later, the tasks mentioned in Table 4.12 were added to the welders' c.q. operators' functions.

In general, Mossink and Peeters conclude, that the 'job content of robot operator's leaves much to be desired' (1992, p. 4). This conclusion leads them to make some recommendations for improvement. Firstly, work content can be improved by interfering in the production and labor organization (not the typical MST-character of this recommendation). Thus, improvement measures should not be restricted to robot-related activities. Programming and optimizing often take little time, and thus offers little opportunity to improve job content. Secondly, the use of buffers can prevent machine-paced jobs (it was observed, that operators tend to keep up with the robot's speed, even when not strictly necessary).

Conclusions and discussion

Unfortunately, the circumstantial nature of the empirical evidence does not allow to draw conclusions with a high degree of certainty. The statistical material was gathered more than ten years ago, in the introduction period of industrial robots. Since then, the number of robots has increased by a factor twenty-three at least (see section 3.2). However, some tentative remarks can be made.

It is necessary to make a distinction between stand-alone robots such as most arc welding robots, and robots which are a part of a larger whole, for instance those at ATAG, Schlumberger, Staalcentrum Roermond, and Volvo\NedCar. About half of the robots used for 'other applications' in the study by Haring and de Mare (1991) are linked to other system components, and perhaps even integrated in a composite system. In such composite systems even pure technical changes can not be attributed solely to robots, let alone consequences for non-technical aspects. As far as the empirical material allows top draw conclusions, a significant part, if not all, of the manual work was automated by the introduction of such composite systems, leading to job losses. Not only the robotized activity itself was automated, but also other activities such as (un-)loading, transport of products and tool setting. The differences between the work around, respectively, stand-alone robots and composite systems probably explain the differences in spending time by operators for the two groups of applications distinguished by Haring and de Mare (1991). (Un-)loading only occurs at the beginning and end of the line, and supervision forms a major part of operator's work, who may also conduct other tasks. As the total production system determines the possibilities to handle a certain product range, i.e. all system components must be able to handle this range, robots incorporated into these systems, are likely to handle a smaller number of different products than stand-alone robots. The implementation of task groups (after years of having used more traditional forms of work organization at Volvo\NedCar) proves that there is a fairly large freedom of choice to form and even experiment with different kinds of work organization around a composite system, in this case the spot welding line, and that this option is actually used.

In case of stand-alone robots, (un-)loading often forms the major part of the operator's jobs. The output characteristics determine the frequency with which these tasks occur. The survey data show that few programs are used and cycle times are often short. Batch sizes vary. Small and medium-sized batches were reported to be most common by Laurentius c.s., but they also report that programs are generally changed only after a few days. The findings of Haring and de Mare (see Table 4.11) also point to relatively long batches, although the average is probably lower for welding robots than for robots used for other applications. These output characteristics probably have much to do with the commonly used teach-in programming method (see section 3.1), which restricts the economic feasibility of handling a large product range (Groover, 1987). Probably, this constraint is more important than the use of variety reducing techniques, which could not be shown to have been applied in connection with robots (see hypothesis 3). Unfortunately, there are no empirical data available on the use of programming methods, although there is a *communis opinio*, that other programming methods than teach-in programming can be regarded as exceptions.

The case of Poutsma shows the effects of the output characteristics on jobs: repetitive and monotonous work is created, which occurs more to the extent that batches are longer and cycle times are smaller. If no or only a small number of new products are introduced, little time is spent on programming. Optimizing and testing only take place if programs are changed, which in 1982 happened once in several days according to the survey data. As case C9 shows, in case of long cycle times the effects of machine-paced work are alleviated. The survey data of Laurentius c.s. and Haring and de Mare, and the cases provide strong indications, that much of the work around stand-alone robots consists of feeding the robot, which regularly is a machine-paced activity. Thus, much work around stand-alone robots is routine. However, if the robots in firm C are confronted with the cases B7-8, D12-13, and the EIM-case, it can be concluded that this is not a necessity, provided that other such as indirect tasks, are added to jobs. One has to realize, however, that firm C's jobs were not improved until worker resistance occurred. These findings confirm the impact output characteristics have on work content as far as operating tasks are concerned. However, hypothesis 2 links the output characteristics to the occurrence of integrated jobs, which relationship is left undiscussed in the empirical material.

In some cases, robots are used to eliminate unattractive jobs, as was the case with monotonous assembly jobs at Schlumberger, and the (un-) loading of furnaces in case A. In 1982, improvement of the working environment was mentioned as an important motive to purchase robots. However, as shown in case A, this improvement is likely to be connected with the solution of recruitment problems rather than to humanistic concerns. Furthermore, such an amelioration seems to be limited to specific applications, mainly handling. Furthermore, as the cases by Mossink and Peeters show, an unsafe working environment and jobs with an unattractive work content may result after the implementation of a robotized workspot. This is not a necessity, and can be counteracted by taking (generally technical) safety measures and applying different design rules to the newly created organizational task pool, as illustrated by the Volvo\NedCar example and Mossink and Peeters. The reorganization process at Volvo\NedCar confirms hypothesis 1 about the independency between organizational design orientation and the technical system used. The same holds for the differences in work content found in the various cases. Unfortunately, no indications were found about the possible importance of the various programming methods for the creation of integrated functions (see section 3.3). A similar conclusion can be drawn with respect to the influence of the labor market, as proposed by hypothesis 4.

4.3 Flexible Manufacturing Systems

The number of flexible manufacturing systems, including flexible manufacturing cells in the Netherlands is small. Given this small number, it is remarkable how many FMSs have been the subject of research concentrating on organizational aspects. Table 4.13 gives an overview of the Dutch FMSs studied, and contains the main sources used.

Table 4.13
List of some Dutch FMSs

Firm	Plant location	Sources
DAF (2 FMSs)	Eindhoven	Krabbendam, Willenborg
Enraf-Nonius	Delft	Lok and Slomp (1990)
Grasso	Den Bosch	Krabbendam, Willenborg; Vrakking (1988)
Humabo	Hengelo	Slomp and Spijkerman (1988); Slomp and Verbeek (1989)
Stork	Boxmeer	Krabbendam, Willenborg; Vrakking (1988)

Although the division of labor does not always form the heart of such research, the studies seem to give a fairly adequate insight into this aspect. Much of this work has been conducted at the University of Twente by Willenborg, Krabbendam, and Slomp. Willenborg and Krabbendam studied six firms using a total of eight FMSs. Three of these firms are Dutch (DAF, Grasso and Stork), the others being, respectively,

Belgian and Scottish. DAF used three FMSs at the time, one of which is installed in Westervoort, Belgium. Consequently, it is left out of this discussion. In the following, Willenborg's 1987 dissertation is used as the major source, but the description is supplemented by data from a summarizing article in English (Willenborg and Krabbendam, 1987).

Table 4.14 contains some characteristics of four Dutch FMSs used by the organizations described above.

Table 4.14
Characteristics of four FMS-cases

Characteristic / Case	DAF-MC	DAF-MO	Grasso	Stork
Plant size (# employees)	5600		400	700
Number of machines	4	16	2	2
Main product on FMS	chassis bow	motor parts	crank house	cubic parts
Batch size	large	large	small	small
Product range	20	16	75	315
Product variability	low	low	low	high
Reduction direct personnel (%)	83	45	40	38
Number of shifts	3	3	2 plus	2 plus
Number of operators per shift	2	8	2	3

Sources: Compiled from Willenborg and Krabbendam, 1987, p 1685 and Willenborg, 1987, p 74 and Appendix 4

DAF produces trucks. In 1987, the firm used two FMSs in its Eindhoven plant, one in the department for mechanical components (MC) and one in the motor department (MO). Both are operated in three shifts. DAF-MC's FMS consists of four machining centres, each with its own tool magazine and pallet carrousel. DAF-MO produces motor parts. The FMS has 16 CNC-machines, which are grouped in a line lay-out. The output characteristics on both FMSs are fairly similar: the product range and variability are low, and batches are large. **Grasso** is the world market leader for compressors. Its FMS is placed in a functional lay-out, although plans existed in 1987 to replace this by a group lay-out. The FMS consists of two separate machining centres coupled to a central computer, and automatic tool and work piece transport. The product range is quite broad, product variability is low, and batches are small. **Stork** Boxmeer produces machinery for textile industries. The FMS has two machining centres, each with a tool carrousel and automatic pallet changer, automatic work piece transport, buffers and loading stations, and a robot brings the tools from the tool magazine to the machining centres and vice versa. According to Vrakking (1988, p. 162), it is unique as it is able to handle 315 different components in batches of 5 to 150 units, and has an average cycle time of 22 minutes. This unique use poses high demands on tool management. A simulation program was used to make unmanned manufacturing possible. The plant has a group lay-out.

Willenborg pays attention to 'organizational policies', which capture among others the notion of organizational design orientations (1987, pp. 77-80). In the beginning of the 1980s, DAF's top management initiated some organizational renewal projects, among which 'flexible production automation', 'quality of working life' and 'total quality control', which were felt to be necessary in order to comply with changing market requirements, especially an increasing number of product variants. The project 'quality of working life' started in 1982, and was a reaction on the 'existing, Tayloristic organization' with its 'strong separation of thinking and doing, rigid automation, many hierarchical layers, and functional departmental structure' (Willenborg, 1987, p. 80). It aimed to reallocate tasks from staff departments to line departments, and from higher levels to lower levels, thereby upgrading the work of production workers. Stork's policy is aimed at the lowering of costs and an increasing 'production flexibility'. The firm is an early adopter of new technical systems. Flexibility and simplicity are the central principles to structure the organization. According to the management, the production process can never be fully planned from a distance, and unexpected events always occur. Because of this low predictability, the policy is to give shop-floor personnel as much decision latitude as possible. Furthermore, one strives not to split jobs in order to have a workforce, which is broadly deployable. Grasso has a hierarchical culture. According to its management, automated machines require a much stricter and more disciplined organization than conventional machines. Consequently, work preparation has to be regulated in a systematic way, and standard ways of working have to be laid down in instructions, which are made by technical specialists. As much as possible, work preparation and production planning have to be regulated ex ante in specialized departments. Production personnel have to follow the rules in a disciplined way.

Table 4.15 comprises data concerning the division of labor around these FMSs.

Table 4.15
The division of labor around four Dutch FMSs

Task	DAF-MC	DAF-MO	Grasso	Stork
Operating tasks\ (un-)loading	1	1	1	1
Pre-setting tools	2	2	2	1
Quality control	1	1	4	4
Internal work distribution	1	1	1	1
External work distribution	2	2	4	2
Work preparation\ NC programming	3	3	4	3
Adjusting new methods	3	3	4	3
Testing new products	3	3	4	3
Testing existing batches	1	1	4	-

Source: Adapted from Willenborg, 1987, p 125

This table needs clarification. The numbers refer to the following employees:

1 Internal generalist, i.e. the FMS-operator.
2 Internal specialist, i.e. another employee employed within the manufacturing department (generally a 'system manager').
3 External generalist, i.e. an employee, employed in another department than manufacturing, who has more tasks than the task men-

tioned.

4 External specialist, i.e. an employee, employed in another department than manufacturing, specialized in the tasks mentioned.

There are some differences between the three cases: Grasso is characterized by a high division of labor, whereas Stork's operators perform a broader range of tasks. DAF occupies an intermediate position. The 'organizational policies' are clearly reflected in the division of labor around these FMSs. In none of the cases, including the foreign firms, integrated functions occur. The absence of integrated functions can be explained by a pressure on machine utilization and the specialist character of programming knowledge. The capital intensity of FMSs leads to a heavy pressure on its utilization. This stress results in a full utilization of the FMS during production hours,and a large number of operation hours, regularly a minimum of two shifts plus some hours of unmanned production (Stork and Grasso) or even three shifts (DAF). Programming demands specialist knowledge, partly because many factors which are not directly related to manufacturing such as the availability of proper fixtures and tools, have to be taken into account to use the FMS efficiently. Thus, in comparison with conventional machines and CNC-machines, work preparation becomes more complex after the introduction of an FMS. In all cases, optimization of programs is part of a specialist programmer's job, although at DAF and Stork the FMS-operators sometimes make simple adjustments in programs for new products.

The degree to which the FMS can perform the distinguished activities automatically is seen as an important element by Krabbendam. This determines to an important degree the elimination of work, and therewith job displacement, all organizations aimed at, and succeeded in producing with less direct personnel (see Table 4.14). Positioning, and the major part of transport are automated. Krabbendam (1988) states, that FMS-operators in general have two important tasks, namely (un)loading work pieces and controlling tool quality, which is necessary because precision requirements are high. This also holds for loading and unloading, which make up the bulk of an FMS-operator's work. (Un)loading requires accurateness, but is a routine job. Nevertheless, all three firms require a qualification level from FMS-operators that is higher than the level requireed from operators of conventional machines, i.e. from LTS to LTS\MTS. Grasso employs two MTS-graduates, who train two LTS-graduates as FMS-operators. Stork recruited MTS-graduates to operate the FMS, who were to cooperate with internally selected CNC-operators with LTS. DAF's policy, at that time, was to recruit MTS-graduates, who were to follow an extensive internal training program. Willenborg explains these rising qualification requirements because of an increased complexity of the operating tasks.

DAF and Stork signal that specialist knowledge is required for work programmers and programmers, but this has not resulted in rising qualification requirements, which is in contrast to the situation at Grasso.

Overall, many direct tasks are automated, and new specialized tasks are created. In comparison with the pre-introduction period, the average skill level rises because much direct work is eliminated, and specialized work is created. This changed organizational task structure is reflected in the new function structure.

The Humabo-case provides another example. Humabo is a machine-building firm, employing 100 persons. Its FMC has two identical machining centres, ten pallets, automatic work piece and tool transport by means of a handling robot, a tool magazine, and parallel programming. It processes 12 different aluminium products, each in approximately ten variants. Batch sizes range from 10 to 200, and have declined since the introduction of the FMC by a factor 3 to 4. The pallets can hold different work pieces, and the cycle time per pallet ranges from 30 to 60 minutes.

The FMC-crew consists of three persons, a system controller and two operators. The FMC is operated in two shifts, with one operator in each shift. The system controller works day-shifts, and conducts the non-routine activities such as programming and testing. He is the only FMC-worker who took maintenance courses, next to operating courses, which makes him difficult to replace. The operators perform routine tasks. They load products on, and unload products from pallets at the loading station, replace worn-off tools, start new jobs, remove chips, and supply lubrication oil. At night, the FMC is operated unmanned. The operator in the evening shift prepares pallets, which are stored in a buffer to be processed during the night.

The more complex and more easily changeable situation initially led to more intensive contacts between the planning department and the manufacturing. More than a year after the installation of the FMC, it was decided to decentralize detailed production planning to the manufacturing department to the system controller. A specialized production planning and control (PPC) system was developed for this purpose (Slomp and Zijm, 1992).

In the article on Enraf-Nonius, which merged in February 1990 with

Oldelft to form Delft Instruments, (Lok and Slomp, 1990), another organizational option is described. Its FMC has two identical machining centres, both with an automatic pallet changer, and two loading stations. The FMC is used for 26 products, under which cast-iron components of the firm's main product, level meters. The number of products handled by the FMS was expected to increase. Batch sizes vary from 25 to 275, with an average of 40. This average was 110 before the FMC was implemented.

Several (classical) sociotechnical principles were worked out in the Enraf-Nonius case. In 1988, the lay-out of component production was changed from functional to 'more group-wise', which transition process was still in progress at the time the article was written. 'Relatively autonomous units' were (to be) created, which are combinations of men and machines which are able to process several products. The FMS is situated in one of these so-called 'production units', named PU-2, which fulfils a model function. Enraf-Nonius had a policy which stressed working independently with delegation and decentralization of authorities and responsibilities. As a consequence of this policy, PU-2 is responsible for production, machine maintenance, testing NC-programs, tool management, and quality control. Within the unit, three functions exist, namely 'unit co-operators' (four persons), 'tool setter' (one), and 'unit manager' (one).

The operators (un-)load the pallets, monitor the production process, and change the tools. They are held responsible for the completion of the detailed production schedule. This schedule is made by the PU-2's unit manager, who also acts as a contact person to other departments. As in the Humabo-case, a tailor-made PPC-system was developed (Slomp and Gupta, 1992) and tested (Slomp and Gaalman, 1992). The operators and the tool setter are consulted before the detailed production schedule is made up by the unit manager. Slomp and Gaalman mention as a point of criticism after the evaluation, that the operators themselves should have a certain latitude to assign jobs to machines (1992, p. 664). The setter takes charge of setting tools, and is responsible for the presence of the right set of tools and their quality. He only works the day-shift.

As in the Humabo-case, at the end of the second shift pallets are prepared for unmanned production during the night. At the time the article was written, possibilities to increase the number of nightly production hours by increasing the number of pallets were being investigated.

Conclusions and discussion

FMC\FMSs are technically advanced manufacturing systems. The machining centres and large number of automated activities such as tool transport and handling, make these systems fairly expensive. This puts a high pressure on FMS utilization with respect to both hourly output and the number of operating hours. In all cases, there are three shifts or a two-shift system is combined with at least three hours of unmanned manufacturing after the second shift. The planning situation is complex, as numerous conditions have to be taken into account: machining capacities, pallet capacities, and the availability of tools. In case of Enraf-Nonius and Humabo, specialized PPC-systems were developed.

No explicit measure was included in the studies to measure product complexity. Probably there is some variation with respect to product complexity: Humabo's cast-iron components may be more simple than the motor parts produced by DAF-MO's FMS. Nevertheless, one may conclude from the case data that the average product on an FMS is fairly complex compared to the products made on conventional machines, which is logical considering that FMSs can combine various operations in one production run. There is also some variation with respect to product variability: in some cases, the configuration is set-up to produce a fixed product range (DAF, Grasso), in other cases new products are frequently introduced on the FMS (Stork, Enraf-Nonius). The product range may also be limited (DAF, Enraf-Nonius) or very diverse (Stork). A similar picture emerges for batch sizes: small (Grasso, Stork, Humabo, Enraf-Nonius) and large (DAF) batches occur, although the latter are a minority in these specific cases. With respect to hypothesis 2, this broad variety on output characteristics makes no difference for the number of integrated functions: they probably do not exist around FMSs in the organizations discussed above, and were not found in other studies (ISI-ISF, 1987; Willenborg, 1990). This can be explained by two reasons. Firstly, the pressure to constantly utilize the FMS means, that (un-)loading is a continuous process. This work is not directly machine-paced, but decoupled from the processing operations by buffers. These guarantee a constant flow of products c.q. a high utilization rate. (Un-)loading takes even more time in case of unmanned operation during the night. In this case, the pallets have to loaded at the end of the last manned shift and the processed work has to be unloaded during the beginning of the first manned shift on the following day. Furthermore, FMS-operators need to monitor the production process, and take appropriate actions in case of unforeseen events. In most cases, the operators are at least partially responsible for scheduling, which is one of the main operating tasks determining the system's productivity.

Secondly, programming FMSs is a highly complex activity, as many factors have to be taken into account simultaneously. This work is done by specialists, which is also true for most other tasks (see Table 4.15). DAF has partially-integrated functions, but in the other cases even optimizing is generally not done by the FMS-operator.

Even though the FMS-operators are not or only hardly involved in programming activities, there is some variation with respect to the assignment of indirect tasks (Table 4.15). Changing, replacing or pre-setting tools and\or quality control may be included in the function of the FMS-operator, but the main part of their work consists of (un-)loading and process monitoring. The overall picture that emerges is one of polarization: routine tasks are conducted by FMS-operators, while specialists take care of the non-routine tasks. Willenborg concludes that the sharp dichotomy concerning the division of labor around the FMSs, which is signalled in international literature, is not present in the cases included in his study (1987, pp. 126-130). A possible explanation might be, that the reduction of direct work is so drastic, that there are only a few persons left which makes an extreme division of labor impossible. Five out of six cases have a maximum of three operators per shift, which leaves little room to create specialist functions for routine activities. Non-routine activities, especially in work preparation, are often assigned to specialists, also because the FMS has to run smoothly, which requires a perfect preparation. The influence of an organizational design orientation is most clear in the assignment of tasks to persons in different departments (see Table 4.15) rather than in the jobs of FMS-operators. Perhaps this is caused by the high pressure on FMS-utilization, leaving little room for large variations in the operator's job.

The Humabo and Enraf-Nonius cases show, that operators work two shifts, whereas specialists works only day-shifts. Operators probably have to be able to work independently during the second shift, although no mention is made of the consequences of the different manning during both shifts for the activities carried out. If FMS-operators conduct only routine activities, there may be little variation between both shifts, but if a specialist, such as Enraf-Nonius' setter, just works the day-shifts, variation may be inevitable.

Discussing Swedish experiences, Bengtsson (1992, p. 38) addresses the issue of 'machine system knowledge', i.e. knowledge concerning the functioning of the system, which is necessary to optimize the daily operation process, solving disturbances, and deal with unexpected events.

Such knowledge is 'particularly insistent in highly automated FMS' in connection with 'continuous operation and availability'. Probably, machine system knowledge is a useful concept to describe the changes in the required qualification levels of FMS-operators. As in process industries, less manual and more intellectual skills are needed. Manual intervention is replaced by computerized control systems. Yet, close process supervision is necessary in order to detect potential disastrous faults in an early stage. FMS machining experience is needed to be able to build-up such knowledge. This kind of knowledge stands in contrast with the bulk of routine work conducted by the same operators.

Detailed labor market data are missing. Combining estimations about the number of Dutch FMS\FMCs and the average number of operators per FMS\FMC, one can conclude, that the total number of FMS-operators in the Netherlands is most probably lower than 200. As well qualified and loyal personnel is generally selected to work with these systems, who subsequently seem to occupy a privileged position, work on FMS\FMCs may be firm or system specific, no general education exists for FMS-operators, and the geographical mobility is limited in the Netherlands, the impact of the labor market is probably negligible except for the clear division reflected in the polarized function structure.

The tendency to raise the educational level of FMS-operators compared to conventional machinists may seem strange if one only takes the complexity of the operating tasks into account, but may be explained by the necessary machine system knowledge or a tendency to entrust capital-intensive equipment to persons with higher educational levels. Willenborg's citation of a manager illustrates, that such a policy may be hard to maintain: 'How can I keep a MTS-er motivated if he has to spend one-third of his time on loading?' (1987, p. 129).

4.4 Conclusions and discussion

Well before the 1990s, the conclusion that technological determinism does not exist, but that there are degrees of freedom with respect to organizational design instead, has become a truism. Thus, a mere reconfirmation of these facts as found in the previous sections, is not sufficient. An attempt should be made to identify variables that influence the choice for a specific design, or put differently, how the degrees of freedom are used. Passing the station of technological determinism, it can be concluded from the empirical material in this chapter that organizational design principles, technical characteristics, output characteristics, and labor market considerations influence such a choice, often in complex interconnected ways. Economic considerations form the backbone of this process, as the choice for any design is influenced by ex ante perceived and estimated economic advantages and disadvantages, and ultimately by the actual economic performance. In the following, these general statements are made more concrete by using the empirical material with respect to automation of discrete manufacturing processes in the Netherlands. For this purpose, a distinction is made between the stand-alone machines CNC and robots on the one hand, and composite systems such as FMSs and an occasional robotized system on the other hand. The latter are generally more capital-intensive than the former, which puts a high pressure on their utilization. This shows in the continuous operation of composite systems during a large number of hours. Both have consequences for the division of labor.

Striving toward continuous operation is an insurmountable obstacle for the creation of integrated functions around composite systems, as these have to be (un-)loaded and monitored continuously, leaving no time for programming. The solution of alternating physical places to allow operators to program, as Peters recommends for CNC-machines, has not occurred, probably because programming composite systems not only requires programming skills, but also insight into many technical and organizational factors, which jointly set the conditions in which the programs have to work. Therefore, the differences in work organizations around composite systems probably show more in other, indirect functions than the operator's function.

The case is totally different for stand-alone machines, where programming and optimizing may or may not be part of the operator's job. Furthermore, various decisions were taken with respect to the division of indirect tasks. The decision whether or not to create integrated functions depends on a broad spectrum of considerations. Less integrated functions occur with an increasing organizational size, an increasing number of CNC-machines, three-dimensional operations compared to two-dimensional operations, and higher programming comfort. Yet, a multitude of other factors, some of which are included in Figure 4.1, play a less evident role, and the relationships between these factors confuse the total picture. A finding of a more speculative nature is, that integrated functions do not occur more, and probably even less, around industrial robots than around CNC-machines, which is discussed further on.

Fruytier and ten Have (1990) and Mossink and Peeters (1992) suggest looking for changes in the division of indirect labor to improve the jobs of, respectively, CNC-operators and robot operators. Such changes concern the job decision latitude of these operators, for instance making decisions on work distribution, work sequencing, the choice of tools and machines, and programming. Including programming in operators' jobs is then only one, albeit important aspect allowing operators a certain job decision latitude. It raises the question to what extent operators can be craftsmen even when they do not program. Is an integrated function around a parametrically programmed CNC-machine more or less 'craftsman like' than the split job of a FMS-operator with a high job decision latitude and full responsibility for the system? Such a question can only be answered after thorough theoretical discussions and the subsequent development of a measure to weigh importance of these different tasks. The empirical material does not provide sufficient data to get insight into the division of indirect tasks, or even of tasks influencing the job decision latitude other than programming. An indicative answer is given by Bell (1990), who conducted research to improve work content at DAF-MO, which was also included in Willenborg and Krabbendam's research on FMSs as well as reserach on CNC-machines (Peters, 1986; COB\SER, 1988). The FMS was left out of his project 'Redivision of Tasks' because 'the FMS-operators already have relatively broad tasks with a high job decision latitude and high job demands' (Bell, 1990, p. 53). Furthermore, the machine system knowledge of independently operating FMS-operators must not be underestimated, and may form the prime reason why their education level is on average higher than that of respectively CNC- and robot operators. Yet, this stands in sharp contrast with the fact, that their work consists mainly of loading and unloading (cfr. Willenborg, 1987, p. 161).

Integrated functions probably occur less around robots than around CNC-machines because of a combination of technical and economic considerations. The relatively simple teach-in method used for robots does not stimulate the creation of integrated functions compared to the more abstract ways of programming CNC-machines. The reverse could have been expected, as simpler programming methods can be more easily learned by operators. The output characteristics form a moderator variable in this relationship. Because of the on-line programming method robots produce, on average, longer batches of fewer products than CNC-machines (also see Hagedoorn, 1988). These different output characteristics have consequences for the work to be performed; the bulk of the operator's work consists of loading and unloading (Poutsma et al., 1987; Mossink and Peeters, 1992).

This last conclusion can also be drawn for composite systems (Vrakking, 1988; Krabbendam, 1988). Both robots and composite systems are, for a combination of technical and economic reasons, more limited than CNC-machines with respect to the product range handled. Although robots are technically reprogrammable, this option may hardly be used because, economically, it is only feasible when a substantial number of products will be produced (according to the expectations). Composite systems are limited in their 'flexibility' because all system components have to be able to handle a certain product range. Writing a program does not suffice, but an aspect like the presence of sufficient quantities of the proper tools and fixtures has to be taken into account as well. Such techno-economic considerations influence the characteristics of the output produced on technical systems in general, and therewith the frequency with which activities occur. For instance, the teach-in method poses strict limits on the economic feasibility of writing new programs, which leads to a low product variability, and hence product variety. Consequently, the programming and optimizing frequencies are low, leaving operating, i.e. mostly (un-)loading, as the foremost important activity in quantitative terms. The frequencies with which these tasks occur put constraints on possibilities to redesign work: for both CNC-machines and robots it was argued, that the inclusion of programming into the operator's job adds little to the operator's work content if programming hardly ever occurs. Furthermore, jobs without a sufficient work content are unattractive for qualified personnel, and are likely to lead to motivation problems.

Although CNC-machines are less restricted in dealing with a large product range, it has to be noted, that this option is not always used. Other technical features than this particular feature, such as the speed of operation and exact reproducibility of workpieces, may just as well have been the prime reason to purchase a CNC-machine. The recurrent observation of a low programming frequency (Peters, 1986; COB\SER, 1988) provides the evidence. The actual use of this option has consequences for the output characteristics and thus the frequency with which tasks occur.

The labor market for technicians with vocational training, whether LTS or MTS, is tight, a situation which is likely to continu in the near future (FME, 1992). The impact of this scarcity is probably higher for CNC-machines than for industrial robots and composite systems. The latter are used less and a substantial proportion of the robots are used for applications which do not require a technical education such as handling and assembly (see section 3.2). Furthermore, FMS-jobs may have a higher status than CNC-jobs, although a fair comparison fully depends on the job contents of both operators (see above). Yet, operators of composite systems need machine system knowledge to deal with unforeseen circumstances. The necessary mental skills will probably be associated with a good work attitude, i.e. a disciplined way of working, and high requirements on the education level, i.e. MTS or well qualified LTS. Whether or not such a policy leads to factual underutilization can not be concluded from the empirical data.

For stand-alone machines, much depends on:

a the assignment of programming, and to a smaller extent of optimizing, to operators. For the case of CNC-machines, this is reflected both in qualification and in education levels (see Table 4.9);
b the division of indirect tasks.

The persistence of the policy to recruit MTS-ers for integrated functions, or as specialized programmers, will form an obstacle for the diffusion of integrated functions.

In conclusion, the dominant coalition has many options to interfere in the process of work and organizational design. Different design principles may be used in connection with decisions concerning product markets, standardization of components, purchasing and using technical systems, and labor market policies. The complex interconnections between these aspects make it highly unlikely that they are all taken into account simultaneously. Furthermore, it must be realized that changes with respect to one aspect have consequences for all other aspects. Such decisions are likely to be partial, a fact which is 'discovered' only after the decision has been taken.

Although the degrees of freedom in work design may suggest a move towards integrative designs, they can just as well be filled in by choosing a fractionized design. *Options are by definition optional. The use of these options is critical.* The assertion, that NC made division of labor possible (Noble, 1978) is just as useful as the assertion, that flexible automation offers possibilities to improve work content (COB\SER, 1984). Literally, both statements point to options, but at the same time they suggest what the direction of the development will be. Dankbaar makes a similar observation with respect to integrative designs (1988).

The application of integrative and fractionized design methods may at the workspot level be influenced by technical considerations, for instance the creation of possibilities to increase the division of labor and programming methods, but this influence has to be seen in the first place as the setting of the conditions within which design principles are applied. Within this decision space, the same factors may even have contradictory influences (see Figure 4.1).

Note

1. The degree of difficulty of this work must not be overestimated, as stated eloquently by an interviewed Volvo employee: 'even monkeys can do that work' (Roobeek, 1987, p. 91).

5 The division of labor around arc welding robots

As concluded in section 4.2, at the beginning of the empirical phase little empirical research on the division of labor around industrial robots was known, compared to other forms of discrete manufacturing automation in the Netherlands. The most common application of industrial robots in the Netherlands, and in many other countries as well (Karlsson, 1991), is arc welding. Arc welding robots are practically without exceptions stand-alone machines. Furthermore, the scarcity of skilled welders, who are necessary to program arc welding robots, provides an opportunity to assess the role of this factor. For these reasons, it was decided to focus on arc welding robots in the empirical part of this study. The body is formed by ten case-studies on the division of labor around arc welding robots.

The cases are briefly described in section 5.2, although some additional facts are given later because of the readability. Before proceeding to these descriptions, methodological considerations are given in section 5.1. The analysis is the subject of section 5.3. The central research question as posed in chapter 1 is answered here. The closing section 5.4 contains the conclusions and discussion.

5.1 Methodology

This section aims to give an overview of methodological aspects concerning the empirical phase of this research project. It should provide the reader with an adequate idea concerning the execution of the project and its limitations. Furthermore, some practical hints for conducting research may be derived from it.

The main part of the empirical study is formed by the cases, although a limited number of interviews were conducted. These concerned persons with some kind of expertise in the field of robotics. Before proceeding to a discussion of methodological aspects of case studies, brief attention is paid to the interviews.

As concluded in chapter four, at the time of the beginning of the empirical phase of this research project the field of robotics was a relatively unexplored area. Therefor, a limited number of unstructured and open interviews were conducted with persons with expertise on particular aspects of robotics. The existence of publications or several citations, both in journals in the field of mechanical engineering, was used as an indication for the expertise of the interviewees. They were working in either the academic or practitioners' world. The content of the questions was mainly dependent on the field of expertise of the interviewee. No explicit research strategy was used in conducting these interviews. Rather, interviews were held during the entire research project if felt necessary and appropriate at the time. To ensure the reliability of this phase, a draft report was sent to the interviewee in order to check its correctness, and adapted if necessary.

The interviews had three goals:

1 To get up-to-date information concerning the state-of-the-art in the field of (arc welding) robotics in the Netherlands. This information was used for various purposes, for instance to find suitable organizations to conduct case-studies and to validate certain findings (see section 5.3). Furthermore, it had a certain influence in the process of model building, but the exact effect here is hard to track down precisely.

2 To try to gain insight into the matter what cases and particular aspects in these cases could be expected to be common respectively exceptional. Probably the most classical objection against the use of case studies is their limited statistical generalizability. Thus, interviews with experts were used to get a better founded idea what practices are common respectively exceptional. An example was the opinion expressed by all interviewees, that teach-in programming is by far the most dominant programming method for arc welding robots.

3 To fill some gaps in the available material concerning particular topics, for instance the information concerning the use of robotics within Philips (section 3.2) and information on the use of off-line programming in the Netherlands.

An advantage of case-studies above quantitative research is, that the researcher gains more insight into the sometimes intricate pattern of occurring relationships than would have been possible with quantitative research (see section 4.1 for an example). Yet, qualitative research, among which case-studies are included, seem to be undervalued compared to quantitative research, perhaps more in the United States than in Europe. Yin states that 'Too many people are drawn to the case study because they believe it is an "easy" method. [...] No perception could be farther from the truth' (1989, pp. 61-62). In his view the absence of standard data collection procedures makes it necessary for every researcher to pay explicit attention to methodological aspects. In the following some methodological aspects are addressed as far as is relevant to case studies as conducted in this study.

Formulating a theoretical model One of the first and most important steps in conducting any type of research, whether using case studies or any other research strategy, is to explicitly define the problem to be researched, preferably in the form of a theoretical model and research questions. Chapters 1 and 2 report on the problem definition and the final theoretical model. Concerning the process of the formulation of this model, it needs to be mentioned that several earlier drafts exist, based on literature and the interviews mentioned. These were presented on scientific conferences (Benders, 1991b; Benders and de Haan, 1991 and 1992) and published as a discussion paper. This discussion paper was distributed to some sixty persons (among whom most interviewees) who could be expected to be able to give adequate feedback on the model discussed. Feedback and comments were processed in the revised drafts and ultimately in the model and hypotheses as formulated in section 2.5 and 3.3.

Validity and reliability For any type of empirical research, four tests are relevant:

1 Construct validity: establishing correct operational measures for the concepts being studied.
2 Internal validity (in case of causal and explanatory studies): establishing causal relationships, whereby certain conditions are shown to lead to certian outcomes, as distinguished from spurious relationships.
3 External validity: establishing the domain to which a study's findings can be generalized.
4 Reliability: demonstrating that the operations of a study - such as data collection procedures - can be repeated with the same results.

Yin (1989, p. 41) discusses case-study tactics to deal with these tests. Many of these tactics were used if felt appropriate, as described further on. Before proceeding, however, the issue of the external validity of cases needs a more elaborate discussion, as it is easily confused with the external validity of quantitative studies. Yin warns about two different meanings of external validity, namely statistical generalizability and analytical generalizability. Statistical generalizability means that a specific sample is representative for a larger population. Statistical generalizability can not be achieved by case studies. No set of cases is likely to deal satisfactorily with statistical generalizability. 'In analytical generalization, the investigator is striving to generalize a particular set of results to some broader theory' (Yin, 1989, p. 44). Case studies can confirm or falsify theories, which points to the dominant role of theory c.q. model building in case study research. Yin recommends the use of replication logic, which is only possible in multiple-case design. This replication logic was warranted by using the same checklist for all case studies, which leads to uniformity of data collection (Kommers and Maas, 1987, p. 14).

In the underlying research, construct validity is of relevance at two separate levels of analysis, namely the measurement of items within cases and the overall case report. Most items are fairly easy to measure; for instance the length of the weekly operating period can be measured by the number of operating hours per week. In order to achieve construct validity for the overall case report, draft reports were reviewed by key informants in the organization studied, as recommended by Yin. Furthermore, if possible multiple sources of evidence were used, most notably documents, interviews with several persons and on-site observation.

Internal validity is important in establishing causal and explanatory relationships, whereby possible spurious correlations are controlled. Strictly taken, the underlying study is not causal, because multi-moment observations are necessary to establish causality. Instead, the broader 'explanatory' seems to be a better description, but the problem of one-moment shots remains. Thus, in a methodological sense the level of certainty about the established relationships does not exceed that of firm plausibility. The 'firmness' is based on:

- the preceding review of theoretical (chapter two) and empirical literature (chapter four);
- two German studies (Urban and Hölldampf, 1988; Lauenstein, Seitz and Volkholz, 1989), which were found after the case material had

been gathered, confirming major parts of the analyses;
- the logic consistency of the relationships.

A problem in finding a general pattern of relationships in the cases is the lack of a standardized mode of analysis. The theoretical propositions, partly in the form of hypotheses, provided a framework to look for specific relationships. A pattern-matching logic was used, interpreting the cases in the light of the propositions.

External validity To achieve analytical generalizability, cases have to be selected for theoretical reasons. Specific details with respect to cases are given in section 5.2 after the cases have been introduced. The basis for the case selection was a report, published by the consultancy agency ANERTEK (1989), containing the addresses of most Dutch firms and institutions using robots as of December 31, 1988. Section 3.2 contains the limitations of this report's data. As this report is the most complete known to be available and statistical generalizability can not be achieved by case-studies anyway, these limitations do not seem to constitute a problem. Two cases (Meubel and Component) are not included in the ANERTEK report, but were found thanks to informants. As some aspects proved to be insufficiently covered by the random selection process, a more selective and theoretical approach was adopted later by looking for specific characteristics (Kommers and Maas, 1987; Eisenhardt, 1989). By including cases that satisfy certain selection criteria, the influence of such factors can be investigated, which is important to test the hypotheses. The cases selected because of such theoretically interesting features are Country and Akker (integrative design), Meubel (size), Plough (number of robots), and Component (programming method). Yet, as other relevant criteria may or may not vary simultaneously with the selection criterion, the evidence has an indicative rather than a definite nature, in other words: this selection process does not fully comply with the strict criteria for quasi-experimental designs (Campbell and Stanley, 1963).

Six organizations refused their cooperation to the study. Reasons that were given were (the number of firms mentioning the reason is given between parentheses):

- the robot was no longer being used due to technical deficiencies (one);
- malfunctioning of the robot (one);
- off-line programming (which would have been the reason to include this particular firm) is not yet operational (one);
- the representative c.q. the organization was too busy (two), respect-

ively 'handling the matter seriously would take more time than is available' (one).

Some plausibility that the selected cases are not exceptional may be the best that can be achieved, which was one of the goals in conducting interviews.

A checklist was formulated in order to guarantee the uniformity in the data collection procedures and therewith the reliability. The case reports were stored in a data base. The most relevant data are described in the following section. The case descriptions in section 5.2 are included to allow others to validate the results, or alternatively, to interpret the material to draw their own conclusions. Where possible, these descriptions contain literary translations of the original statements as made by the respondents, recorded in case-study reports, and later verified by the respondents themselves. The use of literary statements fits with the idea of design orientations, in which perceptions of dominant individuals form arguments to make decisions, f.i. about the division of labor.

Data collection After the first contact had been made with a knowledgeable representative of the organization to be researched, a copy of the checklist was mailed to this representative. This procedure allowed the representative to get a better idea of what the research was about and what efforts were required from the organization. When the organization agreed to cooperate, the checklist was filled in as far as possible from written documentation in order to collect the data as efficient as possible. An additional advantage of using written documentation is that it is often a precise source for (technical) details, which interviewees are not likely to recollect at first hand. Although the documentation was not always up-to-date (organigrams are notorious in this respect), the reviewing of the draft report by the key informant warrants the reliability of the data.

Most of the other data were collected during on site visits by means of interviews with informants, such as chefs of department, personnel managers, robot programmers and operators. In most organizations the actual robot workspot was visited as well.

Some cases have also been covered by other research projects, namely Plough and Akker (Peters, 1986) and Falcon (Mossink and Peeters, 1992), although Peters only dealt with CNC-machines.

Research team Some cases were executed by graduate students, who wrote their master's theses on this basis. These cases are Yellow (Leermakers, 1991) and Iron, Falcon and Dies (van der Vegt, 1992). As there are no standardized ways of conducting case studies, all researchers have to be fully aware of the goal and limitations of the project. Furthermore, they need an adequate background knowledge. Thus, both students were involved in respectively the formulation of a theoretical model, and in the critical assessment of a preliminary draft of the model formulated.

5.2 Case descriptions

In this section, brief descriptions of the individual cases are given to allow readers to make their own analyses and make this study's interpretation controllable. After these case descriptions, a cross-case analysis is made in section 5.3. The organizations are first introduced in a general way, and briefly discussed. After that, more detailed data concerning the division of labor are presented.

Yellow Yellow's three divisions produce various metal products. The organization employs about 630 persons. Its world-wide sales were approximately Dfl. 180 million in 1991 with a 10 per cent profit margin. The firm has a strong market position: it is market leader in the Netherlands, and belongs to the top five European manufacturers. Its strategy is customer and quality oriented.

The visited plant produces canalization systems, and employs 150 persons. It has two manufacturing departments, namely 'standard components' and 'technical systems'. The canalization systems are composed from standard components. Three robots are used at 'standard components', namely one spot welding robot and two arc welding robots.

Country Country produces systems for the internal transport of discrete products. These are ordered by, and designed in cooperation with, specific customers, for which project teams are formed. Country produces and installs these tailor-made transport systems. The organization has 400 employees. Its 1990 sales were Dfl. 160 million with a profit margin of about 6 per cent. The Dutch market accounted for 15 per cent of total sales; 73 per cent was sold in other EC countries, while the remaining 12 per cent was exported outside the EC.

Its only plant is located at the headquarters, at which site 320 people work: 128 in manufacturing and 192 in non-manufacturing departments. The manufacturing department has four manufacturing units (construction, sheet metal working, turning, and assembly) which are all located in one large hall. Country uses one arc welding robot.

Aerie Aerie produces agricultural machines, pumps and cleaning equipment, mainly for professional users. Its 90 employees generated total sales of Dfl. 16 million with a 5 per cent profit margin in 1990. It is the Dutch market leader, and exports mainly within the EC. The image is that of a high quality and innovative firm. Aerie's products are modular: by combining different components, 16,000 variants of its main product are possible. These machines are often built according to customer specifications: specific wishes, which can not be fulfilled within the existing range of components, are investigated with respect to their technical and economic feasibility.

Its only plant is located in the same building as the headquarters. There are three manufacturing units in the functionally organized plant. The arc welding robot is placed in the construction unit.

Plough Plough is a part of the division 'agricultural machinery' of a large German enterprise. Its 500 plus employees generated total sales of Dfl. 160 million in 1990, but losses were made because of the EC agricultural policy, which leads farmers to postpone their investments. Therefore, Plough decided to temporarily abandon its second shift. The machines are standard products, but Plough holds some important patents for internally developed product innovations.

Two manufacturing plants are located close to the headquarters. The visited plant has 180 employees, most of whom work in the two main manufacturing units, preparatory processes and assembly. Three arc welding robots are placed in the sub-unit welding.

Meubel The main products of Meubel are wooden and metal chairs and tables. These are sold to intermediate traders and institutions, for instance in health care. Meubel also acts as a subcontractor for large manufacturers of furniture, who can not produce smaller quantities of non-standard products. Its market share is small, and its profitable sales were Dfl. 5,5 million in 1990. The firm employs 17 persons (including both directors), eleven in the manufacturing department and six in other 'departments'. One arc welding robot is used in manufacturing.

Akker This organization became independent from Plough after a management buy-out in 1988. As a part of this deal, it was agreed to slowly decrease the level of fixed orders from its former mother plant, necessitating Akker to develop new markets. At the time of the visit (November 1991), the economic situation was bad, and Akker accepted most orders it could get. The firm acts as a jobber as well as a sub-contractor for various customers. Its market share is small. The 1991 sales were about Dfl. 12 million, leaving a small profit. The two-shift system was temporarily abandoned, and only 60 out of the regular 90 employees remained. Of the remaining 60 employees, 48 work in manufacturing. The welding unit uses two arc welding robots.

Petra Petra is part of a so-called 'social workplace', a government-sponsored, non-profit institute to employ and revalidate 'those who are not able to take part in the ordinary labor process'. Similar organizations exist in all regions of the Netherlands, and were founded to execute the 'Bill of Social Work Provision'. The organization makes a broad range of generally fairly simple products, some on order, others at own risk. Its sales of Dfl. 2,8 million were generated by approximately 180 cooperators. Most cooperators conduct manual assembly operations. The manufacturing unit 'metal', in which two identical arc welding robots (called Peter and Petra) are placed, has 25 employees.

Iron and Falcon Iron and Falcon are two manufacturing plants of a Dutch enterprise, which has six manufacturing sites in Europe. It produces a broad range of quality household appliances, which are exported to other European countries. Its current strategic policy aims to increase its market share in other European markets by offering well-designed and modern products. The organization has over 1,000 employees, and total sales in 1991 of Dfl. 168 million with a profit margin of three per cent.

The 143 employees of site 'Iron' manufacture four different household products, and perform a range of administrative and sales services. One arc welding robot is used in the tube department.

Falcon's 253 employees produce a broad range of small products. The plant was destroyed by fire in 1989. Following this disaster, two arc welding robots were purchased in 1990. At the time of the visit in 1991, the plant was again fully operational.

Component Component forms a part of a large Dutch machine-building enterprise. This organization produces various kinds of machinery, and exports world-wide. The specific division of which Component is part, has a fifty per cent share of the world market. Its high price is compensated by the reliability and innovativeness of its products and its good service.

Component makes various parts of manufacturing plants, which are sold as a whole by the mother company. It receives its orders on-line from the mother plant located at the divisional headquarters. The 55 employees with fixed contracts are normally assisted by approximately 15 persons with agency contracts. Together they generate total sales of Dfl. 15 million. Component has one arc welding robot, which is programmed by means of a hybrid off-line programming system.

Table 5.1 summarizes basic case characteristics as described above.

Table 5.1
Basic case characteristics

Case	Plant size	Branch (SBI)	# robots	Year
Yellow	150	34.8	2	1986 '88
Country	320	35.4	1	1987
Aerie	90	35.1	1	1983
Plough	180	35.1	3	1979 '86 '90
Meubel	17	34.5	1	1990
Akker	90	34.8	2	1981 '89
Petra	180	39.5	2	1988 '90
Iron	143	34.8	1	1989
Falcon	253	34.8	2	1990 (2x)
Component	70	35.3	1	1987

According to the CBS (1991, p. 46), 18,0 per cent of all Dutch robots were used in organizations with less than 20 employees, while organizations employing more than 99 persons used 55,9 per cent. The remaining 26,1 per cent was employed by organizations in the intermediate size category. All these categories are represented in the underlying case-studies, although the smallest size category contains only one organization.

Nine organizations are in the metal products (SBI-codes 34) or machine-building (SBI-codes 35), which can be explained by the particular field of application, namely arc welding. The remaining organization Petra (SBI-code 39.5) is a non-profit institute for social work, also producing metal products.

All organizations with the possible exception of Meubel had resolved the initial (technical) problems, and robot operation had become a routine, standard activity.

Except for one case, all organizations are located in the South of the Netherlands (eight in the province of Brabant; one in Limburg). This region is a traditional stronghold of metallurgic industries, which partly explains its innovativeness as shown in the diffusion data of section 3.2. This geographical delineation reduces the influence of a possibly disturbing factor, namely different business conditions in separate regions. Personnel shortages are less problematic in the South than in the rest of the country: 77 per cent of FME-members in the Rhine delta report recruiting difficulties, 50 per cent face similar problems in the South. The national average was 56 per cent (FME, 1990c, p. 44). Yet, labor market conditions are not homogeneous within the region.

Country and Akker were known to have adopted integrative organization design principles. Country had already been visited in February 1990 on another occasion, and Akker is included in Peters (1986), when Akker had plans to reorganize. Petra was included because of its specific situation with respect to the capabilities of its employees, and its elaborate and conscious project approach used to make the investment decision and subsequent implementation of the robots. Petra's spokesmen presented their approach at a conference.

One visit was paid to an organization, called **Dies** here, with 66 employees producing metal products, which sold its robot after two years of production for three reasons. In the first place, the robot was not really suited for the thin metal parts (diameter of 0,5 to 3 mm), which had to be welded. Secondly, the average batch size of the firm's products were too small to use a robot. In the third place, the robot had been acquired to produce large batches of some components, which had been ordered by two customers. When these orders were cancelled, the robot was sold in 1989.

The robot had been used to produce three products in batches ranging from 100 to 1500. New products were not introduced, as described above, and the robot was not felt to be suited for welding the thin parts. Probably, the fact that the robot had been out of production because of technical malfunctioning for the larger part of the two years, has also played a role in the decision to sell it.

The division of labor

Yellow uses three robots, one for spot welding, the other two for arc welding. Both are operated for 45 hours per week. The manufacturing department has a functional organization, grouping machines of the same type in 'clusters', which are each headed by a 'cluster leader'. The spot welding robot is placed in an other cluster than the arc welding robots, which are placed in a 'general cluster', that contains machines for various operations. Arc welding was a new joining operation for Yellow, for which the robots were bought; the spot welding robot replaced manual welders. The robots were programmed by a HTS-educated employee of 'work preparation', but after that his programming capacity is only addressed when more complex optimizations have to be carried out. Occasionally agency temporaries are hired to (un-)load the robots in case of large batches, but they are monitored by a regular robot operator. The work of robot operators includes (un-)loading, optimizing, setting-up, quality control and minor repairs and maintenance. They need to have welding knowledge. Both the cluster leader and the programmer often assist with more complex optimizations. This activity occurs quite frequently with 2 to 10 product changes daily. Batch sizes vary from 50 to 4000 pieces. Both robots make variants of one simple product, with respectively 23 and 49 variants.

Decisions concerning the division of labor were only made after the robot had been purchased. Yellow's current policy is that machine operators have to learn to operate all machines within their cluster. 'First-class machine operators' are required to operate at least one extra machine, which takes about one year and makes it easy to replace employees in case of absenteeism. 'Operation' here includes setting and optimizing programs. The cluster leader is able to operate all machines in his cluster, which is to become the norm for all operators. The robot operator's skills are valued higher than those of the operators of conventional machines; the conventional machines are prepared for production by specialized setters, whereas the robot operators do the setting themselves. Yellow's CNC-operators, however, need more operational knowledge and have to change tools, which requires more skills than the work of the robot operators.

Most machine operators have completed an LTS-education. This was felt to be too low to create integrated functions, whereby the operator has to program. MTS-educated operators were not available, but the same choices would have been made if they had been available, because MTS-ers would not have had sufficient career perspectives even with integrated functions.

Country bought its robot in 1987 to replace part of the manual welding. It is placed in 'construction', and operated in two shifts, i.e. 78 hours per week. The robot operators have integrated functions. They set the machine, and control product quality. The maximum batch size is 400, and about 45 products are produced. Programming occurs every week.
If the product is not welded properly by the robot, this is corrected manually by the robot operators. There were two reasons to choose for integrated functions:

1 Coordination losses which occur with the other options are prevented.
2 The work is kept attractive for LTS-ers, which are hard to recruit.

Country also has integrated functions around its CNC-machines, with the exception of some older machines which are controlled by means of punch tapes. These punch tapes are made by work preparation, but this practice will end as soon as these machines are replaced by more modern equivalents.

Originally, the robot operators were either construction mechanics or welders. These are LTS-educated employees with welding knowledge. Country's construction mechanics get work drawings from work preparation. These drawings do not contain specifications of welding parameters: the mechanics have to decide on their own how to construct the product. Given the large product range and variability, their work can be classified as craft work. The same holds for the robot operators, who have to make their own decisions regarding the welding programs.

Initially, two 'better' mechanics were selected to take care of the start-up phase of robotic welding, which was then a new application for Country. When the process became routine, they were replaced by two other mechanics, whose construction capabilities were felt to be below average.

Aerie's robot is also placed in its functionally organized 'construction' department, where nine manual welders work. The robot is operated during day-shifts only. Its capacity equals two to two and a half manual welders. Initially, the robot was acquired to weld large batches of light frames, i.e. fairly simple products, for which Saudi Arabia was an important market. When this large export order was cancelled, Aerie was faced with the problem how to keep the robot productive. It had not been foreseen, that the robot might have to be used for other products. This drawback led to a significant fall in the robot's productivity, which was ultimately resolved. Currently, fairly large and complex products with a total welding time of up to one hour, in small batches (between 25 and 100) are processed by the robot. Sizeable products and small batches are welded manually. The changed output characteristics necessitated additional investments in manipulators, welding tables, and fixtures. The latter are made internally by the programmer and three other employees.

Aerie employs two robot operators and one programmer. By alternation, the operators work at the robot or weld manually. They perform all robot-related tasks with the exception of programming. Many welded products need finishing, which is done by the robot operator or manual welders. The total workload, generated by finishing, is too high for the robot operator. Setting, quality control, material supply, and cleaning also belong to the function of the operators. The programmer occupies a special position, although he does not form a separate hierarchical level. In general, he performs more difficult activities such as maintenance and making fixtures, which can take up to two weeks. The division of labor between programmer and operators is not a sharp one, and becomes increasingly blurred. The operators are already entitled to make minor adjustments in programs, and on a longer term one strives to teach the operators how to program, but the loss of production time forms an obstacle. A situation with more programmers would be preferable, because the dependency on only one person would be reduced.

The main reason for the current situation with divided responsibilities is the labor market situation. Welders are hard to recruit, which is even stronger true for programmers. The latter need to possess welding knowledge, but they also have to be able to think in an abstract way. Given the tight labor market, such programmers are in the position to pose high demands on their work content: 'A competent programmer will resign if he has to load [a robot]'. The momentary programmer is known to have had a job offer, which he turned down because he would have to (un-)load the robot in that job. At Aerie, the work around the robot is not considered to be attractive for everybody, and that operators may also resign if they are not content with their work. Currently, the one more capable operator experiences the work as challenging, which has much to do with the fact that the work is not yet planned perfectly. Many adjustments have to be made, for instance due to variations in the parts to be welded.

Plough was among the first robot users in the Netherlands. In 1979, its first arc welding robot was installed, which was part of this suppliers' first shipment of robots to the Netherlands. Later two arc welding robots were bought from another brand. The latter are more user-friendly, because they are specialized arc welding robots whereas the first robot is a general-purpose robot applied for arc welding. This user-friendliness was, next to the larger working domain, an important reason to opt for the other brand. All three robots are placed in the welding unit, where automated and manual welding take place as well. All capital-intensive equipment, including the robots, is operated in two shifts, which consist of three persons, namely two operators and one operator-programmer.

The operators have split functions, and keep production records. All optimizations and programming on the three robots is conducted by the operator-programmer (called 'programmer' from now on), who also performs the operating tasks. The operators are unskilled employees. The programmers are skilled LTS-educated welders, who were selected from the (internal pool of) manual welders because they were perceived to be better skilled and were interested in operating the robot. This was more or less a promotion, because they were given responsibility for all robotized welding processes. The high repeat frequency, and thus repeated optimizations, are an important element to keep these jobs attractive for the programmers. Furthermore, about once every month a new product is added to the existing range of 45. Batches vary from 50 to 1000 pieces.

Plough would prefer to have operators with integrated functions, i.e. who are able to deal with all tasks around the robot, but this option is currently not feasible because of a lack of skilled welders. Unskilled employees are less hard to recruit than skilled welders. This problem is worsened by the fact that the robots are worked in two shifts (during regular conditions). Welders are not always prepared to work two shifts.

The small firm **Meubel** bought its arc welding robot in December, 1990. It was placed in the manufacturing hall, where eleven persons are employed in three different areas: metal working, furnishing, and assembly. The production workers may be deployed in any of these areas, depending on the workload.

Batches vary from 100 to 300 pieces. Currently, the welding of some 20 different products is robotized, to which product range a new product is added about once every two months. Meubel has split functions. The operating tasks are conducted by the production worker who happens to be available. One of the firm's managers occasionally works with the robot himself. A punch machine has to be operated simultaneously. There is a certain variety (varying products on different machines) in the operators' work. This is possible because this work is unskilled, and can be performed by any 'normally talented' person. Programming and optimizing are done by an LTS-educated programmer-foreman, who is also responsible for work distribution and setting. Furthermore, he makes the welding fixtures, and conducts occasional repairs and maintenance. Yet, he is not considered to form a separate hierarchical layer. He was chosen for this work because of his welding capacities and the fact, that he had shown to be ambitious and willing to bear responsibility. He would probably leave if he had to operate the robot. Meubel is dependent on its only experienced programmer. He could probably be replaced on a short term, were he to leave the organization for some reason or another. Meubel's other welder has taken the robot course together with the current programmer and the manager, but only the programmer has sufficient working experience. This programmer would probably leave the organization if he had to operate the robot.

Akker has two arc welding robots of different brands in use, that weld respectively light and heavy products. Akker's first robot was bought in 1981 for one product, which was too complicated to be welded by an automatic welding machine. Later, other products had to be introduced to keep the robot productive. Currently, the product range of both robots is 25 products, with batches of 100 to 500 pieces. New products are frequently introduced, i.e. once every two or three weeks. During ordinary business conditions these are operated in two shifts. Due to a lack of orders there was only one shift at the moment of the investigation.
In November 1985 Akker started to reorganize. The functional lay-out was replaced by a group lay-out. In 1991, however, the largest sub-department of 22 persons was still divided into three groups, but both robots were placed in a (functional) welding sub-unit.

Akker employs four operator-programmers, who all have integrated functions. However, they are specialized on either one of the robots. They also conduct a broad range of tasks, which are not robot-related, such as setting, quality control, material transport, keeping production records, and all maintenance. The programs are written based on work drawings, provided by work preparation. There were two reasons for this option:

1 A division of labor with separate programmers and (non programming) operators would lead to constant quarrels between both,

because the latter would have to fix the faults made by the former.
2 The robot operators have to be able to solve problems, which occur during the night shifts, on their own. This calls for employees, who can control all facets of the process.

Work distribution is part of the group leader's job. The fixtures are made internally by the specialized tool unit, where nine persons work.

The robot operators were selected from the internal pool of employees. They have an LTS-education, supplemented by on-the-job experience and welding diploma's. Selection criteria were affinity with electronics, welding competence, and the fact, that they dared to operate this new equipment.

Recruiting welders is not much of a problem in the rural area where Akker is located. When necessary, they could be hired from employment agencies in close-by cities. The fact that Akker used to employ more welders when it had more orders in its portfolio, may also have played a role. The recruitment of shift workers, however, was more problematic.

Petra bought its first robot in 1988, which was followed by an almost identical machine in the beginning of 1990. The machines have different welding guns, because the brand used on the first robot was not available any more in 1990. The fixtures and the programs are not interchangeable for this reason. Both are used to weld the same product, which happens in two steps. Each step takes about three minutes. This product proved to be a market success, leading to a quickly rising and high demand. Nevertheless, both machines are still operated in one shift as Petra has very restricted possibilities to recruit personnel. Being a social workplace, potential employees have to be chosen from a waiting list. This list consist of persons, who underwent a medical examination to assess their capability to work. Because welders are hard to recruit anyway, and Petra's first goal is to re-validate its employees in order to return to regular firms, its necessary welding capacity had to be secured, in this case by purchasing robots.

Four robot operators conduct all routine robot-related tasks by alternation, and take care of material supply and of the transport of welded products. The 'work leader', head of the manufacturing unit, takes care of organizational tasks, i.e. work distribution, keeping records, as well as quality control. The operators are unskilled. Skilled welders would not accept this repetitive work (mainly loading and unloading). Yet, for the unskilled operators Petra's revalidation goal can be achieved by gradually increasing the operators' work content. Job rotation, learning to diagnose and restore disturbances, changing the welding wire, cleaning and preventive maintenance are possible measures. Their functions are perceived as upgrading by these operators.

The organization's chief mechanic wrote and optimized the programs for both robots, which was a unique event: only one product is welded, which is identical for both robots. The chief mechanic also made the fixtures, which activity could be closely coordinated with programming by having them done by the same person. He is responsible for setting and maintenance. The robot-related work is only a small proportion of the jobs of the work leader and the chief mechanic.

At **Iron**, the robot is used in the tube department, which has 16 employees. The organization is functional. The robot is used to weld three light products, namely two kinds of frames and a protection rod, in batches of 2500 to 4000 pieces. Both frames are the main products, and are used to produce thirteen different versions of an end product. The robot is operated in one shift because the demand for frames is not yet high enough; its utilization rate increased from 50 per cent in 1989 to about 75 per cent in 1991.

The robot operator conducts all operating tasks and optimization as well as quality checking and some maintenance. He used to be a CO_2-welder. He is the only person with sufficient working knowledge of the robot. When the robot is not in operation, he is given other work or in extreme cases a day off. Programming hardly ever occurs. The robot was pre-programmed by its supplier, and if additional programs are needed in the future, the supplier will be called. 'Programming is contracted out'.

Iron's sister **Falcon** uses two arc welding robots, which replaced two automatic welding machines. Both robots are used as specialized machines: one produces two frames in batches of about 4000 pieces, the other welds one frame (average batch size 2000) and, occasionally, another component (about 3000 annually). The three frames are used in 79 versions of the end product. Both machines are practically constantly used, and the utilization rate is over 90 per cent. The organizational unit where the robots are placed has two shifts, although occasionally a third shift is necessary to handle peaks in the workload. Every shift consists of fifteen persons, of whom two robot operators. The operators conduct all robot-related tasks except for programming. This activity has not been necessary in the last couple of years, because the product range has not undergone any changes. The initial programs were provided by the robot supplier.

The robot operators are either unskilled or have an LTS-education. All have welding knowledge.

Until 1988, **Component** was functionally organized. This structure has been changed to a group structure. The restructuring started with the newly formed group or 'cell' in which the robot is placed. The group has one arc welding robot, one computer-controlled bending machine and a conventional press, which are necessary to produce the 30 different hooks (batches vary from 600 to 900 pieces). Once every four months new products are introduced, generally other hooks. Component tries to to have other products welded by the robot as well.

The design of these hooks varies substantially. Yet, this variety was reduced before the cell was implemented because the bending machine could not handle certain hooks.

The idea to reorganize came from a student from the Eindhoven University of Technology, who held an internship at the time. The reorganization was felt to be necessary because of two reasons. In the first place, there were logistical reasons. The lead times were too long, and were brought back from three months to one week by reorganizing. This led to a reduction of inventories of fifty per cent. In the second place, many craftsmen who worked with Component, started to find their work monotonous after a while, which had a negative influence on their motivation. In the new situation, the craftsmen are responsible for making complete components rather than only conducting a small number of specific operations. More tasks are nowadays part of their job content, and their wages were raised. New employees have to agree to take about five years of additional schooling and to work the existing two-shift system. The low absenteeism, between four or five per cent, is viewed as an indication of the reorganization's success.

The employees of Component enjoy a good reputation at the regional labor market, although labor turnover is not higher than anywhere else. One tries to keep turnover low by making work content attractive rather than by increasing wages. Employees who resign often get non-manufacturing jobs, for instance in work preparation. Asked about possible scarcities of welders, Component's spokesman answered: 'Nothing of the sort is known to us'. The inflow of students at the regional schooling institutes was felt to be adequate.

There are six manufacturing groups, which are headed by a group leader. The group leader does not form a separate hierarchical layer, but is an experienced craftsman who conducts organizing tasks and assist when problems occur next to working in production as the other group members. The group size varies from 3 to 12 persons, and all groups work two shifts.

The group within which the robot is placed and hence called 'robot cell' in the following, has two regular employees, the 'cell operators', and currently two employees hired from an employment agency. Each shift consists of one cell operator and one temporary employee, hired from an agency. The cell operators were selected after informing who was willing to work within the robot cell, which was at that time a new and experimental part of Component. Two inexperienced, but well motivated LTS-educated employees were selected and sent to a course, given by the robot supplier. Both operators have integrated functions, and perform all occurring tasks within the cell. These include quality control, setting machines, internal work distribution, simple preventive maintenance, and registering production data. All three machines are operated by the cell operators. The same integrative design principles are applied in the rest of the organization. Yet, there is also a division of labor within the robot cell. The agency temporaries only perform routine activities, i.e. loading and unloading the machines, monitoring the production process and quality control, and material supply. They also switch between machines as a result of a policy of task rotation. Yet, all preparatory and other non-routine tasks are conducted by the cell operators.

5.3 Cross-case analysis

Having described the individual cases, it is possible to conduct a cross-case analysis with respect to work design. As in chapter 4, the categorization of Alders and Christis (1988a) is used to analyze the robot-related tasks. Thus, the following categories of operators' functions are distinguished:

- I Integrated function (I), including programming, optimizing and operating.
- II Partially-integrated function (PI), including optimizing and operating.
- III Split function (S), including only operating.

First, the division of labor is discussed as a separate item. After that, the influence of the output characteristics, the technical c.q. programming system, the labor market and some other factors as well as their interrelationships are discussed.

Table 5.2 summarizes the case data concerning the division of labor.

Table 5.2
The division of labor around the arc welding robots

Case	Option	# indirect tasks	Remarks
Yellow	PI, PI	3	Progr. frequency is zero
Country	I	2	
Aerie	PI	1	
Plough	S, S, I	1 (2)	Only one programmer
Meubel	S	2	
Akker	I, I	5	
Petra	S, S	1	Progr. frequency is zero
Iron	PI	2	Progr. frequency is zero
Falcon	PI, PI	2	Progr. frequency is zero
Component	I	6	Includ. work distribution

As shown in Table 5.2, many different options were chosen with respect to the division of labor, which holds for both the direct and indirect tasks.

Three organizations can be categorized as 'integrative', having opted for integrated functions. Although Plough has one integrated function in every shift operating its three robots, the organization can not be considered integrative, because the possibility to split jobs, offered by its number of robots, has been used to the fullest extent. The three pure integrative organizations named the following reasons for choosing this option: other options than the integrated option might lead to losses caused by coordination difficulties (Country and Akker), the operator has to work independently during night shifts and must thus be able to conduct all occurring tasks (Akker), the adoption of an organization-wide integrative approach to decrease lead times (Component), and attractiveness of the operator's function (Country) which has consequences for his motivation (Component).

The organizations with split functions named the following arguments. The programmer would probably leave if he had to operate the robot (Meubel), 'qualified welders would not accept the repetitive operating work' (Petra), and the shortage of qualified workers in combination with shift work (Plough). In this last case, the organization would have preferred to have integrated functions, but this was felt to be infeasible as qualified welders are hard to recruit, especially if they have to work shifts. Unskilled operators for the split functions are less hard to recruit than qualified welders.

Confronting the reasons to choose integrated functions with those to opt for split functions, one sees that organizational concerns predominate where integrated functions were created, whereas arguments referring to the labor market are exclusively mentioned in case of split functions. The organizational concerns include the prevention of coordination difficulties, independently working operators, and the robot operator's work in the perspective of the overall organization (re-)design. Besides these pure organizational concerns, the consequences of work content for operator's motivation were mentioned. Where split functions occur, the privileged position of qualified welders is always mentioned as the predominant reason to have created split functions.

With the exception of Aerie, the organizations in which partially-integrated functions occur have only very simple products, requiring occasional simple corrections (Yellow, Iron, Falcon). With Aerie, the tight labor market was mentioned as the main reason not to choose an integrated function: 'a competent programmer will resign if he has to load [a robot]', but on the other hand one tries to gradually teach its two operators to program.

The division of indirect tasks is roughly correlated to the division of direct tasks. The average number of indirect tasks performed by operators rises from 1,2 with split functions, via 2,2 for partially-integrated functions to 4 with integrated functions. Most of these indirect tasks are routine activities, such as visually controlling product quality and supply of materials. Only at Component, the robot operator is involved in a regulating task, namely work distribution. But even in this most integrative organization, the borders of this design have been met: an agency temporary, who only conducts routine tasks, is hired when an increased work load has to be dealt with. Within Component's robot cell, a certain division of labor occurs when the workload increases.

In total 17 robots are used in these 10 organizations. During normal business conditions, the 17 robots are operated in 27 shifts per week. In these shifts, these are manned by 17 operators with partially-integrated or split functions and 12 programmer-operators, i.e. operators with integrated functions (Country has two operater-programmers per shift). Thus, the majority of robot operators in this particular sample can be considered to perform only operating tasks, i.e. unskilled work. The functions of a considerable minority, however, includes skilled tasks.

Resuming it can be concluded, that the work content of most of the operators studied is small, and that they hardly ever have control over their own work. As far as the welding process is concerned, only operators with integrated functions and programmers have process control. Beyond the welding process, only Component's robot operator is allowed to arrange his own work distribution. As two out of the three organizations with integrated functions were explicitly selected for having such functions, it seems unlikely that integrated functions are overrepresented in this sample. When one accepts this last statement, one has to conclude that most work and jobs around arc welding robots have to be classified as routine and unskilled. Only in organizations which explicitly stress the adoption of some kind of integrative design, the work of robot operators is broader. In all other organizations, which lack such organizational policy, the labor market situation is often used as an argument not to choose for integrated functions.

The output characteristics throw more light on the circumstances which favor the creation of integrated functions.

Output characteristics

Before proceeding to the discussion of the influence of the output characteristics, it must be noted that in section 4.1 mention was made of a low programming frequency (Peters, 1986), which was repeated in section 4.2 for robots (Mossink and Peeters, 1992). As noted by these authors, integrated functions have little meaning if programming hardly ever occurs. Because of the on-line programming method, low programming frequencies can be expected to occur more with robots than with CNC-machines. Whereas quantitative research has not taken this aspect into account, it seems useful, especially in case of robots, to include it in this study. However, a modification of Alders and Christis' categorization by including a separate category for cases with a low programming frequency would lead to a contamination between one of the output characteristics, namely product variability which is reflected in the programming frequency, and the organizational options (cfr. Benders, 1990). Therefore, no separate category is created, but this prohibitive factor for the creation of integrated functions is mentioned explicitly. Table 5.3 contains data about the output characteristics. Where two or more robots are used and their output characteristics are not the same, both are included separately.

Table 5.3
Output characteristics

Case	Product range	Batch sizes	Programming frequency
Yellow	1 (23) 1 (49)	50-4000 50-2000	0
Country	45	1-400	once every week
Aerie	45	25-100	once every three months
Plough	45	50-1000	once every month
Meubel	20	100-300	once every two months
Akker	25	100-500	once every two weeks once every three weeks
Petra	1	(1 product)	0
Iron	3	2500-4000	0
Falcon	2 1	3500-4500 2000	0
Component	30	600 - 900	once every four months

Regarding the robot's 'flexibility', which is here understood as 'the ability to be reprogrammed in order to handle variations in the product range', some observations can be made. About half of the robots were bought to produce one specific product or just a small number of products. This occurred at Yellow, Petra, Iron, Falcon, Akker (only its first robot), and Aerie, involving a total of six organizations and nine robots. The first four organizations still use their robots in this rigid way, whereby no product variability occurs at all. At Yellow, Iron and Falcon, the robots are specialized on one or two specific products, and Petra uses two identical robots for one product. Iron and Falcon had the robot supplier deliver the programs as well, a policy which was emphatically rejected at Petra. Petra's robot was purchased after a thorough preparation, during which robot-using firms were visited. It was concluded that some of these machines 'disappeared under the plastic' because their owners had failed to build up robot knowledge internally. At Petra, an important argument to buy robots rather than automatic welding equipment, which seems to be more fit for their situation, was that robots would be able to produce another product in case the current product could not be sold any more. Thus, in Petra's case a conscious decision was taken to buy robots with 'flexibility' as an important argument. Yet, the flexibility is latent in this case, and will only become manifest when a new product (range) is actually introduced.

Akker, Aerie and Dies tried to convert from a small to a large product range. This process generated substantial difficulties, especially because this possibility had not been accounted for beforehand. Dies sold its robot, and Aerie's robot was underutilized for a substantial period. By undergoing a long trial-and-error process, however, Aerie ultimately managed to make a broad range of products, which are unique in the sense of size and cycle times (up to one hour). This requires substantial knowledge about making fixtures. Furthermore, additional investments in computer memory and welding tables c.q. manipulators were made. Akker's reason to buy a robot for one specific product was, that this product could technically only be made by a robot. This particular robot has two welding guns, which makes it more complex to use for other products. Ultimately, Akker's second robot, purchased eight years later than the first, is expected to handle a larger product range than the first one. Whereas Aerie and Akker ultimately succeeded in converting to a broad product range, Plough, Meubel and Component meant to produce a large number of products from the start.

It can be concluded, that robots are used in situations with strongly varying output characteristics. This conclusion is vital to discuss hypothesis 2: 'The lower product complexity, the smaller product variability, the smaller the product range and the larger batch sizes, the less likely it is, that integrated jobs are found'. Roughly, two situations can be distinguished, namely stable versus dynamic output characteristics, as shown in Table 5.4. Note that product complexity is not included, as the nature of its influence is different from that of the other output characteristics.

Table 5.4
Output characteristics and organizational options

<table>
<tr><th>Output characteristics</th><th colspan="2">Stable</th><th colspan="3">Dynamic</th></tr>
<tr><td>Product range
Batch size
Program. freq.</td><td colspan="2">1-3
50-unlimited
0</td><td colspan="3">20-45
1-1000
at least once every four months</td></tr>
<tr><td>Option</td><td>S</td><td>PI</td><td>S</td><td>PI</td><td>I</td></tr>
<tr><td>Organization</td><td>Petra</td><td>Yellow
Iron
Falcon</td><td>Meubel
Plough</td><td>Aerie</td><td>Country
Akker
Component</td></tr>
</table>

According to hypothesis 2, it can be expected that stable output characteristics are coupled to split and partially-integrated functions, and that dynamic output characteristics are linked with integrative functions. Table 5.4 shows that far more combinations occur than is assumed by this hypothesis.

Yellow, Petra, Iron and Falcon fit in the stable category, and the remaining organizations belong to the dynamic category. Switches between categories are possible, as demonstrated by Aerie and Akker, who both switched from the stable to the dynamic category.

The output characteristics influence the frequency with which tasks occur. In case of stable output characteristics, programming never occurs, optimization scarcely (or in the extreme case of Petra never, leaving the split option as the only feasible one), and operation forms the bulk of the work. Thus there are only two alternatives left for operator's functions: partially-integrated (Yellow, Iron, Falcon) or split functions (Petra).

Whereas the number of organizational options is limited in case of stable output characteristics, dynamic output characteristics do not necessarily coincide with other options. The integrated function forms an alternative, but other options may still be chosen. At best, dynamic output conditions provide a necessary, but insufficient environment for the creation of integrated functions. The cases of Aerie, Plough and Meubel show that the favorable circumstances offered by the dynamic output characteristics were not sufficient for the creation of integrated functions. In Plough's case, the number of machines plays an important role (see below). The other firms with dynamic output characteristics (Akker, Country and Component) opted for integrated functions. All of these has opted for an integrative design, not only for the robotized workspot, but for the entire organization.

Many firms seem to strive toward stable output characteristics for its robots. Yet, the impact of variety reducing techniques is modest at best (see hypothesis 3). Only Component reduced its product range, because one of the other machines in the cell where the robot was placed, could not handle the product range. Hence, the product range that can be handled by a group of machines is equal to the range that can be handled by the machine with the smallest handling capability, analogous to the smallest link determining the strength of a chain. Some products out of the organization's total product range are usually selected to be welded by a robot. Typically, these are products which have to be produced in sufficient quantities and are not too complex (see below). Yellow, Country, Aerie, Iron, Falcon, and Component are all examples of organizations, where they use the robots for component production. These components are later used in a large variety of end products. Yellow, Iron and Falcon succeeded in creating such a stable production environment, that the product variability is zero. By striving towards stable output conditions, organizations create an environment which is favorable for fractionized designs. Dies, Aerie and Akker faced the opposite development, namely from an intended situation with stable output characteristics to changing output characteristics, which was caused by external and unexpected circumstances, namely the cancelling of orders. Unlike Dies, Aerie and Akker managed to adapt to this new situation, be it that Aerie faced underutilization during a considerable period.

At Country, another aspect of dynamic output characteristics showed. Country's robot is only used for about 38 per cent of total production time. The remaining time is needed for optimizing and programming (11 per cent), placing fixtures (24 per cent), and gathering material (28 per cent). Although recently a second robot operator was added, which measure was expected to double the utilization grade, this situation shows the effects of a high programming frequency and a high product variety on the utilization grade. The production situation becomes more complex, and more preparatory tasks have to be conducted, which asks for appropriate organizational measures. Otherwise, the robot's capacity is underutilized.

So far, the output characteristic 'product complexity' has not been given attention. Product complexity can be assessed by the product cycle time.

Complex products, consisting of many components, need more welds and at different locations than simple products of light material. Both the number of components, and thus the necessary number of welds, affect the product cycle time positively. Yet, the same holds for the material to be welded: thick plates or 'exotic' metals such as titan require a longer welding time than respectively thin material or ordinary forms fo iron. From a technical point of view, both these factors also increase welding complexity and the welding time. Thus, the cycle time can be considered to be a good operationalization of (a technical conception of) product complexity.

Three categories of product welding times can be distinguished:

I Less than 2 minutes (Yellow, Iron, Falcon).
II Between 2 and 8 minutes (Plough, Meubel, Akker, Petra, Component).
III More than 8 minutes (Country, Aerie).

On average, product complexity is fairly low. Few products need more than five minutes to be welded. With respect to the operator's work, product complexity has an important consequence, namely machine-paced work can occur with short cycle times. At Iron, the cycle time of approximately 30 seconds is even shorter than the set-up time.

The effects of machine-paced work can be mitigated by designing fixtures in such a way, that they contain several products, which has been done at Component. A measure which can reduce the monotony is having different products welded after another. At Plough, different components of the same end product are welded in one cycle. This is possible by placing the individual welding programs in a batch and by using several welding tables and fixtures. This policy allowed Plough to reduce its work-in-process. Meubel also has production runs consisting of various products.

At Petra, the length of the cycle time was an important consideration. The cycle time must be longer than the set-up time to prevent strongly machine-paced work. On the other hand, if this gap is large, operators might get bored.

Technical system

According to hypothesis 1, the organizational design is independent from the technical system(-s) used. For arc welding robots, the use of off-line programming is relevant in this respect. This programming method requires a higher conceptual level than the teach-in method, leading to a larger skill divergence between programming and operating tasks. In this case, one would expect off-line programming to coincide with fractionized designs, i.e. split or partially-integrated functions. Yet, hypothesis 1 is confirmed in the case of arc welding robots. The operator in the only organization using a variant of an hybrid off-line programming, namely Component, had integrated functions and the broadest range of indirect tasks of all operators. The particular programming systems consists of a sort of spreadsheet, which contains the process parameters. However, these programs are not ready to use, and need calibration at the workspot.

Although this finding does not prove that off-line programming is likely to lead to more non-integrated functions, it does point to the fact that the application of a hybrid off-line programming is not an insurmountable barrier to create integrated functions. One may hypothesize that in regular production situations (expensive) off-line programming packages are only bought when they are expected to be used frequently. Otherwise, this investment would not make economic sense. To validate this hypothesis, additional information was gathered by means of telephonic interviews. At least four Dutch organizations have experimented with off-line programming of arc welding robots, but only one organization had succeeded in its implementation. However, this particular welding process was not a standard welding process. In all four cases, there were (intentions to create) integrated functions, which confirms the view, that off-line programming, albeit requiring abstract programming knowledge, does not form a barrier for the creation of integrated functions. Yet, this relationship has a somewhat spurious character. The purchase of an off-line programming package for standard production situations is only economically advantageous if it is used sufficiently, i.e. when the programming frequency is high. All these organizations have dynamic output characteristics, or more specifically, generally produce unique variants of highly complex products. Such variants leave much time to program during the welding process, and substantial welding as well as programming knowledge is needed (cfr. Figure 4.1). Thus, production situations in which off-line programming is an economically attractive alternative also provide ideal circumstances for the creation of (high level) integrated functions.

Another indication affirming hypothesis 1 is the fact that the teach-in method does not guarantee the creation of integrated functions. Obviously, for arc welding robots the influence of other factors than programming methods is stronger than the programming methods themselves. The foremost important reason is the design orientation: Component clearly incorporated the robot in an integrative organization design. Another reason is indirectly connected to technical characteristics.

Next to diverging skill requirements, another important difference between teach-in and off-line programming concerns economic consequences. As repeatedly asserted, teach-in programming constraints the economic feasibility of producing a large product range. None of the organizations produces more than 50 products with its robot(-s). This disadvantage might be solved by replacing teach-in programming by off-line programming. This technical characteristic influences the economic feasibility of dynamic output characteristics. These output characteristics influence on their turn the division of labor, as discussed above. Yet, the economic feasibility is further constrained when different types of fixtures and welding tables are needed. This is generally the case when the products are not just slight variations of each other. Investments in fixtures can be considerable, i.e. up to thirty per cent of the robot system. Especially the smaller organizations with a large product range of strongly diverging products and a high product variability, make these fixtures internally (Aerie, Meubel, Akker) to economize on these investments. Aerie was able to produce a larger product range after purchasing additional manipulators.

Labor market

At some firms, a shortage of welders had been an important motivation to buy the robot(-s), especially at Plough and Petra, and to a lesser extent at Yellow, Aerie and Meubel. In other organizations this argument was only of minor (Country) or no importance (Akker, Iron, Component). Yet, in none of the organizations the introduction of robots led to firing employees. Instead, the formulation is: 'if the robots had not been introduced, and if welders had been available, more welders would be employed to produce the current quantities'. In general, this effect is in the order of 1 to 1,5 additional employee for each robot, i.e. 2 to 2,5 manual welders produce the same output as one robot with its operator. The replacement of manual welding by robots is an important motivation to introduce these machines, but it has to be noted that this effect is rather limited compared with mass mechanization, which may be one factor explaining the slow diffusion of robots (cfr. section 3.2).

The respondents of individual firms had strongly differing perceptions about the labor market scarcity of welders. Akker and Component ('nothing of the sort is known to us') experienced no difficulties attracting welders. However, Akker's position is rather special, as the firm faced a strongly declining demand for its products, resulting in the firing of its less competent welders while retaining the more competent ones. On the other hand, Akker had already opted for integrated functions when its business conditions were still favorable. Probably due to its integrative design orientation, Component enjoys a good reputation at the local labor market, which was said to supply sufficient welders. Other organizations (Aerie, Meubel) stated that they would employ more welders had they been available, or mentioned labor market scarcity as the prime reason to purchase robots (see above).

Educational requirements vary with work content. The case of Petra is a clear confirmation of the unskilled nature of the operating tasks. On the other hand, programmers need welding knowledge as well as knowledge how to operate and program the robot. Most programmers possess a LTS-diploma or a comparable level. Yet, only welding knowledge, preferably certified, is a *conditio sine qua non*. Equally important are less palpable traits such as 'well motivated' and 'interested', which often serve as criteria to select robot programmers out of the internal pool of qualified employees.

With respect to the fourth hypothesis, 'Labor market scarcity at intermediate segments forms an obstacle for the application of integrative designs', the following observations can be made. All organizations with dynamic output characteristics which do not have integrated functions (Aerie, Plough, and Meubel) mentioned labor market scarcity as an important reason for this decision. The organizations with integrated functions (Country, Akker, and Component) mentioned organizational arguments. A possible conclusion is that the labor market scarcity can be overcome by offering integrated functions. In that case, the solution is found in a measure at the demand side rather than reacting to the supply side. These integrated functions prove to be sufficiently attractive for qualified welders, but it has to be mentioned that the integrated functions at Country and Component also contain many indirect tasks. Akker forms an exception to this rule, but as stated this firm occupies a special position experiencing worsened business conditions.

The organizations with stable output characteristics all opted for partially-integrated functions, with the exception of Petra. Petra's employees are less able, and optimizing hardly ever occurs at Petra. This choice can be seen as a contradiction of hypothesis three, because optimizing requires some programming knowledge, and higher skills than the operating tasks. Yet, the products of these organizations (Yellow, Iron, and Falcon) are all simple, having cycle times of 33 seconds at most, which makes optimizing much simpler than with more complex products. Nevertheless, at Yellow the cluster leader and the programmer frequently assist with optimizing. Furthermore, the skill level needed to optimize is substantially lower than the skill level needed to program. This is even stronger the case with the relatively easily programmable robots than for CNC-machines (see chapter 4).

Regarding hypothesis four, it can be concluded that labor market scarcity at intermediate segments is sometimes used as an important argument to opt for other than integrated functions. The same problem can be overcome by creating integrated functions. This has to be possible, i.e. the output characteristics are classified as 'dynamic', and is a result of an conscious integrative design policy.

Other factors

Two other factors influencing the productive capacity need discussion here, namely the number of robots in use and the number of operating hours per week.

No systematic relationships between the division of labor and the number of robots were found. Integrated, partially-integrated and split functions prove to occur with the use of one robot (respectively Country and Component; Aerie and Iron; Meubel) c.q. two robots (respectively Akker; Yellow and Falcon; Petra). With this limited number of robots, no systematic influence could be detected, and it has to be concluded that all combinations are possible and were actually effectuated. Only Plough occupies a special position in this respect, not only because of the number of robots, but also because of its special organizational arrangement. Plough's programmer-operator takes care of all robot-related activities on its three robots, while the other two operators only conduct operating tasks.

The influence of the number of robots was further investigated by means of two telephonic interviews. This proved to be possible when additional information was acquired. These firms are fairly well comparable; the second respondent even spontaneously mentioned the first as its competitor. They not only produce the same kind of product with comparable dynamic output characteristics, but also use the same number and type of robots, be it that one organization employed two additional robots of another brands. Major differences concerned the organizational options and the labor market. Being located in a traditionally economically strong and still growing region, the organization with specialist programmers (and thus separate operators) experienced severe recruiting problems. The organization with integrated functions was located in a region with little industrial activity, and thus few opportunity alternatives. In the latter case, the attractiveness of the robot operator's function was mentioned as the reason to have integrated functions.

The following observations were made with respect to the number of operating hours. In six cases, the robots are currently operated in only one shift, which was also the regular operating period for the rest of the plant. In case of Plough and Akker, which are both heavily dependent on agricultural mechanization, market demand had fallen back so drastically, that the former two-shift system has been (temporarily?) repealed in favor of a 'regular' 40 hour operating week. Since both cases show interesting differences and their regular operating period consists of a two-shift system, they are included in the analysis of cases using a two-shift system. However, these examples do make clear that the number of operating hours is heavily dependent upon business conditions within a certain period, which is to be expected as changing the number of operating hours is generally an easier and reversible way of reducing production capacity than buying or selling physical capacity.

Unmanned production is impossible as long as (un-)loading can not proceed automatically. The mechanization of (un-)loading is technically complex, and thus capital-intensive. Thus, in none of the cases unmanned production occurred as this is not yet technically and economically feasible.

In all cases with an effective operating period of more than 44 hours a week, a two-shift system was used. Table 5.5 provides some data concerning the potential influence of shift work on the division of labor as there may be differences between both shifts (see section 4.1).

From Table 5.5, it can be concluded, that no differences in the division of labor between both shifts occur in the organizations investigated, except for the finding that production planning and other preparatory tasks have to be conducted during the day-shift (Plough), and that the day shift is sometimes preferred to program complex products because of the presence of supportive services (Component).

Table 5.5
Shift work and organizational options

<table>
<tr><th rowspan="2">Organization</th><th rowspan="2">Output characteristics</th><th colspan="2">Organizational option</th></tr>
<tr><th>First shift</th><th>Second shift</th></tr>
<tr><td>Country</td><td rowspan="4">Dynamic</td><td rowspan="3">Integrated</td><td rowspan="3">Integrated</td></tr>
<tr><td>Akker</td></tr>
<tr><td>Component</td></tr>
<tr><td>Plough</td><td>Split and integrated</td><td>Split and integrated</td></tr>
<tr><td>Yellow</td><td rowspan="2">Stable</td><td rowspan="2">Partially-integrated</td><td rowspan="2">Partially-integrated</td></tr>
<tr><td>Falcon</td></tr>
</table>

All organizations with integrated functions work two shifts (during regular business conditions). Akker mentioned the fact that the operator has to act independently during night shifts, which is one of the reasons to opt for integrated functions. Country finds itself in a similar situation. Plough solved this problem by having one programmer-operator for all three robots. This can be viewed as a tentative indication, that an extension of operating hours may have a favorable impact on job design, at least as long as it is technically impossible to have the production process proceed automatically. The remaining two organizations with a two-shift system, Yellow and Falcon, have stable output characteristics, i.e. a routine production process, allowing partially-integrated functions.

Furthermore, the negative impact of working night shifts is overcome in the integrative organizations. Whereas Plough mentioned increased recruiting difficulties for qualified welders because of its two-shift system, the integrative organizations do not seem to be bothered by this.

An interesting relationship between product complexity and the number of operating hours was found at Aerie. This organization manages to extend its effective operating period by four hours a week by having the robot work through the breaks, which is only possible due to the extremely long welding time per product (up to one hour). This relationship between the length of the welding time, or, in general, cycle time and the number of operating hours has not been reported in other research.

The length of the operating period is positively related to the number of robots, as the average number of robots in organizations with a two-shift system is 1.87 compared to 1.25 in organizations working one shift.

Finally, the organizational size was not found to be related to the division of labor. However, a noteworthy observation was made at the small organization Meubel, where the manager occasionally operates the robot himself. One may expect an increasing organizational size to be positively related to the degree of the division of labor. However, such a relationship is more likely to show at the organizational level than at the level of the individual workspot, which formed the focus of this project.

5.4 Conclusions and discussion

Figure 5.1 is a graphical representation of the factors that play a role in the formation of robot operators' jobs, and the interrelationships between these factors. It summarizes the relationships as described in the previous section.

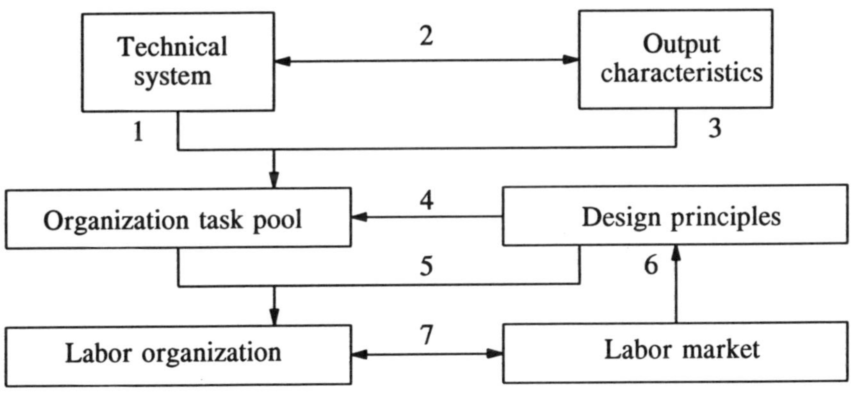

Figure 5.1: The formation of jobs around technical systems

Figure 5.1 is an adaption of Figure 2.2, putting more stress on some specific relationships. Especially the role of the output characteristics is emphasized. The output characteristics make the role of the product market (see Figure 2.2) more concrete. This is possible because in this chapter the level of analysis is the (robotized) workspot whereas Figure 2.2 applies to the (higher) level of the organization. This also possible to link the output characteristics to the technical system under study. On the other hand, the changed level of analysis has its limitations too: the purpose of this study is to investigate the division of labor around arc welding robots, i.e. around a specific technical system. Analogous to the reasoning of chapter two, where the design rules were assigned a more important role than the technical system, it has to be kept in mind here that the introduction of a technical system only affects the organizational task pool, and does not completely determine it. Next to the changed emphasis on output characteristics, a second major change is introduced in Figure 5.1: more explicit consideration is given to the formation of the labor organization as the result of a (changed) organizational task pool and the application of design rules on this organizational task pool.

The 'organizational task pool' is placed in the middle of Figure 5.1. The organizational task pool can be seen as the totality of all tasks that have to be conducted by human beings within the organization. In the first place, the organizational task pool is the result of the totality of products which the organization makes, and which can be described by means of 'output characteristics' (arrow 3). Certain tasks have to be conducted to produce these products, resulting in the organizational task pool. This effect has a qualitative (what tasks have to be conducted?) and a quantitative (how often do these tasks occur?) dimension. In the second place, the organizational task pool is influenced by the technical systems used. In other words, part of the necessary tasks to produce the output is carried out by the technical systems(-s) used (arrow 1). The technical system has several effects on the organizational task pool. Firstly, technical systems generally conduct tasks which otherwise would have been conducted by humans, i.e. the use of technical systems ceteris paribus leads to the elimination of work for humans. Secondly, simultaneously with mechanizing and thus eliminating tasks, machine-related tasks are created. Once installed, the technical system dictates these machine-related tasks: its way of operation is part of the technical system. The frequency with which these tasks occur, however, is determined by the output characteristics. In other words: how the technical system is used in terms of production. There are limits to the output that technical systems produce which have to do with the more or less limited physical capacities and economic feasibility of handling a certain output (arrow 2).

The tasks of the organizational task pool are divided over organization members via design principles, i.e. rules that are consciously or unconsciously applied to decide who has to do what. This process leads to the formation of functions, which can be considered as bundles of tasks to be conducted by individual organization members. Together, they form the labor organization (arrow 5), i.e. the totality of all functions that exist within the organization, for which the term 'function structure' can also be used. The latter term is preferred when labor market aspects are taken into account; the function structure is the organization's demand for employees, which meets the supply of labor at the labor market (arrow 7). As skills are necessary to conduct tasks, job holders have to possess the necessary skills to conduct each individual task, that is part of their jobs. Strongly diverging skills can be an obstacle for the combination of the underlying tasks into functions. The same reasoning applies to tasks created by the introduction of new technical systems: the skill levels required to fulfil these tasks may or may not show variations.

The labor market supply may form a constraint for the formation of functions. In case of scarcity of a particular sort of employees, other design principles may have to be applied to divide the existing organizational task pool in another way over organizational members, in such a way that all tasks will still be fulfilled (arrow 6).

The design principles are concrete manifestations of underlying design orientations, i.e. ideas about the efficiency of a certain extent of dividing labor. In section 2.1 two contrasting design philosophies were distinguished as both extreme poles on a continuous scale, namely the fractionized and the integrative design philosophy. These design principles influence the organizational task pool: the inevitable consequence of dividing labor over more than one person is that the work of these persons has to be coordinated, i.e. coordinating tasks are created (arrow 4). The burden of coordinating activities can be an important reason to choose for integrative designs, which have a lower need for coordination.

The reasoning as depicted in Figure 5.1 can be applied to the technical system 'arc welding robot'. The introduction of arc welding robots leads to the introduction of robot-related tasks such as programming, optimizing, (un-)loading and maintenance. Some of these tasks, for instance maintenance and (un-)loading, may already have formed part of the organizational task pool, as they had to be conducted for other machines, but they are new as far as the robot is concerned. In general, i.e. in the majority of cases where manual welders work, the robot eliminates (part of) the manual welding. Thus, the organizational task pool is affected by the elimination of manual welding and the introduction of robot-related tasks (arrow 1). The frequency with which the latter occur are determined by the output characteristics, or the way the robot is used for production (arrow 3). Programming frequencies range from once a week to once in the robot's lifetime, and the optimizing frequency largely depends on the number of times that one switches to other products. During operation, the bulk of the operator's work consists of (un-)loading, which is interrupted by switching to new products (arrow 2). Although the robot has an aura of 'flexibility', it need not necessarily be used to produce a wide product range. Several firms were found to produce about 45 different products within a certain period, which is perhaps the upper limit. The robots of other firms are used for a highly limited product range, namely not more than three different products, although slight variations in the form of product variants occur. The use of a robot for a broad product range is limited by a combination of technical and economic reasons. As a consequence of the complex kinematical configuration of robots, an overwhelming majority of robots is programmed on-line by means of the teach-in method. The resulting loss of production time is an obstacle to introduce new products: this is only economically useful if sufficient quantities will be produced. The introduction of off-line programming is not likely to change this situation dramatically, as sometimes costly adaptations to the robot system are necessary to produce many different products. Especially welding fixtures and tables c.q. manipulators may be needed. Furthermore, the greater the product range and the differences between these products, the more complex the production situation which has to be controlled.

The tasks of the changed organizational task pool have to be divided over robot operators and other organizational members. In case of dynamic output characteristics, i.e. when programming occurs and there are frequent optimizations due to product changes, organizations may opt to include programming into the operator's function, i.e. apply integrative design rules to create integrated functions. With stable output characteristics the choice is far more restricted as not integrated functions can be formed. The resulting split or partially-integrated functions generally include few indirect tasks (arrow 5). Such an integrative policy is not necessarily restricted to robot-related tasks, but may also take indirect tasks into account. The application of design rules to the organizational task pool leads to the formation of functions, which together form the organization's demand for labor. This demand is confronted with the labor supply at the labor market (arrow 7). In case of a small supply of persons with the needed skills, in this case qualified welders who are needed to program, these persons find themselves in a comfortable position. In some organizations, this labor market scarcity is an important argument to opt for other than integrated functions, leaving the less attractive tasks for unskilled operators. The welders' c.q. programmers' work in these organizations includes programming, and possibly optimizing, next to a multitude of other tasks. In these cases, the labor market was an argument to apply fractionized design rules (arrow 6). Other organizations, however, used integrative design rules, thereby creating jobs that are sufficiently attractive for qualified welders c.q. programmers. In this case, the need for coordination between programmer and operator is eliminated (arrow 4).

Two features of Figure 5.1 deserve special attention. Firstly, the technical system only fulfils a moderate role in the creation of jobs. It affects the organizational task pool, but the formation of the labor organization is a matter of applying design rules to this changed organizational task pool. Both fractionized and integrative design rules can be applied to the same organizational task pool, resulting in totally different work designs. Secondly, there are many loose links in Figure 5.1. These concern the way the robot is used, which has to be seen in connection with the output characteristics and consequently, the chosen approach of the product market, the choice of technical systems and design rules, and connected with this the organization's labor market policy. At such decoupling points, there is a certain degree of choice in making decisions. Different options can be exercised with respect to every single aspect mentioned.

6 Epilogue

'Technology and work' is likely to stay a pervasive theme in scientific as well as popular literature. As the underlying study is part of this stream of literature, it seems appropriate to place the conclusions drawn in the previous chapters in the light of the historical debate of the theme, which is the purpose of this epilogue. Furthermore, in the course of doing so, the accurateness of the robot's friendly image is assessed critically by discussing work content, and hence (required) skill levels, and working conditions in robotized work situations.

Probably technology, or preferably 'the technical system', is given too much attention in discussions and debates about the theme 'technology and work'. This is not surprising, because organizational changes often go along with, or are even induced by, technical changes. The changes with respect to the factor 'labor' occurring after the introduction of a new technical system are easily ascribed to the visible change in the technical system rather than the somewhat impalpable notion of 'organizational design orientation'. The human fascination with technical innovations may be the root of such an interpretation of technical determinism. Yet, as repeatedly asserted in the preceding chapters, the organizational design orientation, possibly concretized in a design method, has to be assigned more weight in analyzing work changes than the technical systems. On the other hand, both aspects are inextricably bound up with each other. A striking similarity between the valuation of changing work content and technical systems concerns the presence of optimistic and pessimistic views on both.

Optimistic and pessimistic views on the development of 'technology' seem to dominate by alternation. There is a widespread belief in the economic advantages to be gained by the development and use of new technology. Even if the opinion about the systems themselves is not optimistic, these economic advantages are a decisive argument to develop and adopt new technology, and form the backbone of organizational, national and international policies stimulating technological developments. A well-known example of a pessimistic vision is the nineteenth century Luddite movement, which set out to destroy machines because jobs were eliminated. Earlier examples are known, such as the banning of weaving machines in Leiden due to weavers' protests in 1621 and the example of Vespasian which formed the beginning sentence of this book. The classical argument underlying a pessimistic view is still a major source of negative attitudes toward new machinery: jobs are lost due to mechanization, and the new work, if any, may be a serious deterioration. However, whereas machines in most cases undeniably take over human work, the impact on the newly created work is in the first place a matter of organizational policies concerning the design of jobs (and work places). There are also optimistic or pessimistic views on the development in the application of design orientations and the consequences of that development for skill levels. In these debates, three different positions are taken, to which (one of) the following labels are attached:

I. Degradation \ downgrading \ deskilling.
II. Regradation \ upgrading \ upskilling.
III. Polarization.

Degradation holds that existing functions are broken up as far as possible, i.e. fractionized design methods are applied. Jobs demanding little skills, often even unskilled or semi-skilled jobs, result. Regradation refers to the opposite phenomenon, namely the integration of tasks. Polarization is more or less a mixture of regradation and degradation: some jobs are enriched, and other jobs are degraded. Polarization is often mentioned in combination with degradation, where the skilled tasks are split of a formerly skilled function and combined into a new skilled job, leaving the semi-skilled or unskilled tasks for others. In this case, many unskilled jobs may be created together with a few skilled jobs, leading to an overall decline of skill levels (f.i. Sorge and Streeck, 1988, pp. 22-23).

At the micro level, the explanations for these changing skill levels can be divided into two major categories, namely the presence of organizational design orientations and the implementation of (new) technical systems. Some authors concentrate on just one category, but others name reasons belonging to both categories. Yet, both causes have a different nature. The implementation of new technical systems affects the tasks that have to be conducted in the organization. Generally, manual work is replaced by one or more machines, which have to be operated. Thus, both regradation and degradation may occur after the introduction of a new technical system. The effects of another way of dividing labor is slightly different: the same tasks are divided in a changed way. As with technical systems, regradation may occur, but the degradation for many is combined with regradation for few, in other words: polarization occurs.

Representatives of degradation, regradation and polarization perspectives have been present as long as the division of labor has been studied. In the following, a historical account of opinions, held by scholars of various countries and political and professional backgrounds, concerning developments in work content and the consequences thereof for skills, and the role played by technical systems in this development, is given. This long-standing debate has become increasingly complex over the years, while at the same time retaining many of the arguments raised in the debate's beginning phases. The emphasis is on manufacturing industries, as these have been given most attention in the historical debate and because the majority of industrial robots are used in manufacturing. In this overview, the terms 'optimistic' and 'pessimistic' are used to describe respectively upgrading and downgrading, thereby following the convention. Yet, it has to be kept in mind, that such these terms almost explicitly contain a normative judgement about the desirable development, and that this norm is humanitarian.

An historical account

Throughout history, the views of degrading respectively regrading seem to have been dominant by alternation, perhaps degradation being dominant. 'The opinion, that the machine, the driving force behind the development of modern firms, and stimulating the splitting up of labor, therewith led to a simplification of the work and a reduction of professional skills, was until recently little contested' (van der Waerden, 1911, p. 1), thus demonstrating that the debate on degradation and regradation is not of a recent date. In his review of literature, the opinions of (political) economists and engineers about the division of labor are discussed. Although these authors often only address particular aspects, and frequently mingle organizational and technical issues, it is worthwhile to reconstruct some of their views and compare these with the opinions uttered after 1945.

According to Smith (1776, pp. I:13-14), the division of labor gave rise to the invention of better working methods, including machines. With the well-known example of the pin-making factory, which he had probably found in an article in a French encyclopedia published in 1755, Smith describes a fractionized labor organization, which by then already was an established way of organizing. Throughout the nineteenth century, the discussion centred on two closely related issues:

1. The social consequences of the fractionized way of organizing.
2. The impact of technical systems on the level of work.

There seems to have been virtually no disagreement on the question, whether or not fractionized designs were being applied, but the impact of new machines was hotly debated. The discussion centred mainly on the social consequences of job losses due to mechanization. Sismondi recommended the use of machines, especially if labor was scarce, but also he also saw the danger that workers, made superfluous by the mechanization, might starve to death (1819, pp. 285-187). Less dramatically, Ricardo also pointed to the unemployment due to mechanization (1817, pp. 264-266). The attention on job displacement is not surprising in an area of occasional revolts against the introduction of new machines, such as the Luddite movement. Around 1800, the replacement of work by machines seems to have been given more attention than the accompanying effects on job content, which is only referred to in a passing way. Smith had already warned about the 'stultification' of work (1776, p. II:303), with which Sismondi agreed (1819, pp. 282-283). All authors agreed on the economic advantages in the form of significant productivity increases to be gained from splitting labor and using new machines (Smith, 1776, p. I:7-14; Sismondi, 1819, pp. 282). Fractionized design methods were constantly being refined, for instance by Gioja and Babbage (1835) (see section 2.1), and the same holds for the invention and improvement of technical systems. De Villeneuve-Bargemont wrote in 1834, that the new machines created unskilled work, an opinion which he thought to be commonly accepted and had earlier been expressed by Sismondi. Yet, some thirty years later Levasseur rejected the opinion that machines led to an increased division of labor and the numbing of man. Instead, they eliminated heavy work.

The discussion seems to have become more animated with the writings of Engels and Marx. 'There is no better means of inducing stupefaction than a period of factory work', according to Engels (1845, p. 179). In the *Manifesto of the Communist Party* it is stated that 'the work of proletarians has lost every independent character due to the growing use of machines and the division of labor', and workers are called 'appendages' of machines (Marx and Engels, 1848, p. 55). Yet, these statements are not representative. In *Das Kapital*, Marx gives detailed accounts of deskilling, drawing among others on reports of the English Labor Inspectorate. In this work, an example is also given of a phenomenon which was later to be called 'polarization', where Marx describes the mass of unskilled workers, often children, whose main task was to feed machines, on the one hand and the 'insignificant number of employees occupied with the control of machines and maintenance' (1867, pp. 402-403). The widely spread unskilled jobs allowed to employ women and children, which lowered labor costs.

The discussion of work content was continued around 1900 by Marshall and Sombart. Marshall incorporated the issue in his discussion of 'industrial organization'. His opinion can be characterized as moderately optimistic. Although he acknowledges, that much unskilled works only exists of feeding machines, ultimately the uniform work 'is sure to be taken over sooner or later by machinery' (1890, p. 255) and new machines require higher skill. Marshall also asserts, that the connection between the improvement of machinery and the growing subdivision of labour is not as close as generally supposed. Instead, the 'increased demand for great numbers of the same kind' leads to the subdivision of labor, and machines can improve certain operations (1890, p. 255). The economic advantages of the combination fractionized organization and new machines is uncontested for Marshall, an opinion shared by the German scholar Sombart. According to the latter, a mutual influence between the division of labor and the application of machines: 'The splitting of labor made machinery possible, as conversely machining technique compels the principle of splitting labor to its utmost consequences' (1902, p. 54). The application of this last principle, and not the use of machines, is the cause of the formation of 'a large mass of purely mechanical activities, which now can be conducted without any problems by the dumbest' (1902, p. 496). The more unskilled work is used, the lower production costs. Furthermore, labor saving is the economic principle of the machine. Sombart makes a distinction between integrating and differentiating machines. In the former category, all activities are mechanized, whereas only part of the activities are taken over by differentiating machines, which are based on splitting labor (1902, p. 531). Another author, Harms, confirmed in 1909 a more subtle view about machines. Harms acknowledged in 1909, that the division of labor led to simplified operations, which lend themselves for mechanization, but he also warned against a biased judgement. Although in many factories mechanization was combined with 'workers who are not more than an automaton', considerable skills were sometimes needed to operate other types of machines. In 1908, the engineer Matschoss expressed the straightforward opinion, that machines took over monotonous work.

The empirical evidence used to demonstrate the described trends was generally very circumstantial, as in case of Smith's pin-making factory, but sometimes elaborate, for instance in *Das Kapital*. The use of casuistic evidence made it possible to demonstrate both degradation and regradation and defend a diversity of opinions concerning the influence of machines. More compelling quantitative data were scarce. An exception to this rule is quoted by van der Waerden, namely the result of a German survey conducted in 1895 and 1907 in commission of the Minister of Trade and Industry. The results showed, that in all but one industrial branch ('art crafts') the proportion of unskilled work had risen from an average of 34.4 per cent in 1895 to 41.0 per cent in 1907. Furthermore, in both years there proved to be substantial differences between branches with proportion of over 90 per cent of skilled work to 15 per cent or less (van der Waerden, 1911, pp. 53-58). His own empirical work, consisting of written responses to his requests for information and factory visits in the Netherlands, showed a similar picture concerning the use of unskilled labor, although no trend could be detected, the project having been conducted only once. Yet, the Dutchman Vliegen asserted in 1908, that 'the demands posed on modern industrial workers are incomparably higher than used to be the case', naming various examples of skilled workers.

Following van der Waerden it can be concluded, that a broad variety of opinions about the development of work content and the related topic of skill levels, and about the influence of technical systems had already been expressed by 1911. Degradation was clearly the dominant perspective and was supported by the German data, but the already existing dissent remained. Another overwhelming impression is, that this degradation process was more or less seen as perhaps regrettable, but certainly inevitable. Manufactures and later factories with an increased division of labor and new machinery outcompeted older production systems. The causes for deskilling were found in the influence of technical systems and the splitting up of jobs. There were several views about the relationship between these two phenomena. According to most authors, mechanization created unskilled, generally machine-paced jobs. Some dissenting opinions were expressed by Levasseur, and later Marshall, Sombart and Matschoss, according to whom unskilled jobs were eligible to be mechanized. Smith's view, that unskilled jobs led to improved working methods found little response, probably because machine-building increasingly became a specialism itself. The dominant view in the nineteenth century clearly was, that machines created unskilled work (Sismondi; de Villeneuve-Bargemont; Marx), which standpoint can be easily explained as technical determinism. Yet, technical innovations were often accompanied by organizational changes, and until Sombart's and perhaps Marshall's work, both causes of the degradation process were generally not separated. Furthermore, only the fractionized design orientation was used, preferably leading the most extensive division of labor possible. The economic superiority of this way of organizing remained uncontested during the entire period and, if any, critical notes concerning the nature of this work were humanistically motivated.

There are several explanations for the differences in opinion. In the first place, there is the circumstantial nature of most of the evidence, which allowed to draw conclusions in both directions. Even though van der Waerden cited quantitative material and conducted empirical research, both of which proved the presence of substantial numbers of unskilled workers, a finding in affirmation with those of most contemporary and at the time already classical authors, counter examples could always be found. The same finding holds for the impact of technical systems. Generally, the machines were fed by unskilled workers, but on the other hand part of the unskilled work was taken over by machines. These were seen to lead to unskilled work (le Villeneuve-Bargemont; Marx), but also as eliminators of heavy and monotonous (Levasseur; Matschoss) and unskilled work (Marshall; Sombart). The varying types of machines (Harms; Sombart) and work around these machines could 'prove' both the degradation and the regradation thesis. Secondly, there is the growth of the number of occupations, often in new industrial branches. On their turn, these may have been downgraded in the course of time, but initially new skilled jobs were created. Thirdly, loose from the factual developments, which as stated above were very hard to map due to the lack of quantitative data and the necessarily impressionistic nature of the remaining empirical evidence, the valuation of these developments differed, and may have influenced the gathering of data. Whereas these academic debates had been present from the beginning on, Marx' work may have functioned as a catalyst. Notably in this respect is van der Waerden's discussion of the views of the prominent Dutch professor and anti-Marxist Treub, who accepted most of the facts and predictions described by Marx, but rejected his interpretation (1911, pp. 11-30). Separate from societal and political viewpoints, the professional background of the engineer Matschoss may be related to his sympathetic attitude toward machines. However, one fact remained uncontested, namely the increasing importance of mass production, or in Marshall's words: 'increased demand for great numbers of the same kind'. Mass production gave rise to and was made possible by an increased division of labor and new machinery. Where craft production remained, it was seen as old-fashioned way of producing which sooner or later was to be replaced by mass production. The economies of scale which could be achieved by means of a fractionized organization and new technical systems must have made an overwhelming impression: even van der Waerden, who stressed the undesirable effects of the deskilling tendency coinciding with the ever-increasing importance of mass production, saw this trend as inevitable, and looked for compensation outside the working situation.

After World War II, much of the discussion was repeated, especially with respect to the supposed degradation and regradation. Especially France formed the stage of a lively debate. Touraine published a study in the Billancourt plant of the French car producer Renault, which formed the basis of his 'three phase theory'. He distinguished three different phases of technical systems, which he connected with specific forms of the division of labor. In the artisanal phase A, autonomously working, professional and skilled working craftsmen make use of several kinds of relatively simple machines. The influence of technical systems increases in phase B's mass production: unskilled workers perform one, or a limited number of highly specialized tasks. Phase B is considered to be a transition to phase C, in which all manual work has been taken over by machines, which are monitored and maintained by new types of craftsmen. A paradoxical situation exists in this last system: 'without a paradox, one must say, that the new system of work, because it is technical, is entirely social' (1955, p. 177). Although Touraine acknowledges the existence of different ways of dividing work in this last phase, dependent on a multitude of social factors, his overall judgement is optimistic. The decomposition of work, which is an essential characteristic of phase B, is necessary in order to have machines take over this work and reach phase C. Touraine places these different phases in a historical perspective, whereby the phases were seen to be dominant respectively in the period before 1880 (phase A), from 1880 to 1930 (phase B), and from 1930 onward (phase C). An extrapolation of this technical evolution to the future leads Touraine to his optimistic point of view, be it, that he cautiously warns: 'these conclusions may not be more than hypotheses' (1955, p. 173). Touraine's standpoint contrasted with that of the influential sociologist Friedmann (de Nijs, 1987, pp. 65-67), who saw a trend of deskilling accompanying the transition from a *milieu natural*, i.e. Touraine's phase A, to a *milieu technique*, which is covered by phases B and C (1946). According to Touraine, however, this evolution may not be judged with the 'arbitrarily chosen ideal' of the traditional craftsman of phase A (1955, p. 14).

In the early 1960s, Mallet placed the discussion in a Marxist perspective, although his empirical basis is not uncontested (see Gallie, 1978, pp. 16-29). With the development of electronics, new industrial sectors were created, which employed 'a new working class' of skilled workers: operators, maintenance workers and technicians. These new workers, indispensable for management because of their firm-specific knowledge, were to play a leading role in overthrowing the existing capitalist society. Another Marxist-oriented French author, Naville, stressed in a 1963 book the disadvantages of work in an automated production environment. Although the trend toward an increasing subdivision of labor was reversed, other new problems in the form of mobility were created. One of these problems was the creation of polyvalent workers, because this policy affected the occupational identity. The existing alienation was accentuated and transformed, not eliminated (Düll, 1975, pp. 194-200).

Meanwhile, in Germany Popitz c.s. published their study of a steel mill in 1957. They also signalled a tendency toward more cooperation and higher responsibility given to production workers in various departments of the steel mill. For the future they expected this tendency to continue in connection with increasing automation (1957, p. 212). In the United States, Kerr c.s. (1960) set out to demonstrate the need for a highly skilled workforce in the future, but this work rests more on a futuristic vision than on empirical data. Earlier, Bright had adopted a dissonant position in *Automation and Management* (1958). He distinguished seventeen different levels of automation, and had studied thirteen organizations with advanced levels of automation. His conclusion concerning skill levels was unequivocal: 'growing automaticity tends to simplify operator duties' (1958, p. 183), although some counteracting effects could be present, such as multi-machine operation and the assignment of tasks as setting-up or inspection to operators' jobs thereby blurring the distinction between direct and indirect work. The same conclusion was drawn with respect to indirect jobs with the exception of some maintenance work. Having concluded this, Bright set out to formulate a 'theory of skill requirements for operators', according to which skill requirements rise in phases two and three, respectively representing the introduction of tools and introduction of power-driven tools, but in further phases the central concept of control was increasingly being taken over by machines, progressively eliminating the need for human skills. Yet, the trend toward growing automaticity will not completely eliminate the unskilled jobs, leaving only skilled jobs in maintenance, set-up and design. In many cases, 'man is a far cheaper and flexible control device than feedback mechanisms for many machines' (1958, p. 188): many activities will be left for humans. Furthermore, '[t]he mix of parts, products, and models required by many sales situations is astonishing', which 'will not support completely automatic machinery'. In addition to that, machines are continuously being improved, thereby reducing the need for maintenance and making maintenance more simple.

Bright's arguments were not sufficiently convincing for everybody. In 1964, almost ten years after Touraine had introduced his three phase theory, the American Blauner introduced basically the same idea as the 'U-curve'. His book *Alienation and Freedom* reports about an empirical study in organizations in the printing, textile, automobile and chemical branches, all presenting stages of technical development. The chemical industry, being highly automated, was held to be 'valuable in pointing to possible future changes in the balance between freedom and alienation' (1964, p. 167). Based on the observations in the chemical industry, where skilled maintenance work and relaxed production work came into existence, Blauner saw the same relationship between the stages of technical development and the skill levels as Touraine, be it that he made some reservations. As Bright has already observed, not all production processes could be automated. Furthermore, some unskilled work would be likely to remain, even in the chemical industry. Finally, the economic situation of the US chemical industry at the time was favorable, but less favorable future circumstances might affect the working conditions negatively. Yet, the overall conclusion was that unskilled jobs were at the time being eliminated at a faster rate than they were created through fractionized design a tendency which could be expected to continue.

Some conclusions can be drawn about the literature of this period. In the 1950s and early 1960s, technical systems were generally envisioned to eliminate unskilled tasks, leaving skilled tasks for the remaining operators. Hitherto, unskilled work had been created, but in the future new machines would eliminate unskilled work. The same period saw the birth of integrative methods of work design, although their success was probably very limited (see section 2.1) and fractionized methods based on Scientific Management were still being developed and refined (f.i. Maynard, Stegemerten and Schwab, 1948). Yet, empirical data showed, that unskilled work was still a common phenomenon (Davis, Canter and Hoffmann, 1955) and the fractionized design orientation still had an overwhelming influence. It is hardly surprising, that the elimination of unskilled work due to the introduction of 'automation' rather than the application of integrative design methods were seen as the driving force behind the coming regradation process. The close association between technical systems and work organization or skills, which is found in Touraine's three phase theory and in Bright's work, can bc explained by the dominance of fractionized designs.

After the mid 1960s, the mood changes. Although Bell (1973) echoed the optimistic views of earlier times, more pessimistic views started to gain popularity. The study of Kern and Schumann (1970) became a classic of German industrial sociology. Their study is based on case studies in nine organizations, operating in eight different industries (the automobile industry is represented twice). A total of twenty organizational changes were studied; nineteen of these involved technical changes, one concerned the reorganization of assembly work, leading to a stronger division of labor and simpler work. Their main conclusion is, that the function structure is more conservative than the degree of mechanization. The latter only partially affects the former, whereby the function structure of the non-affected part is kept intact. Furthermore, unskilled tasks and subsequently functions repeatedly re-occur after the introduction of new technical systems, for instance when manual work is replaced by repetitive 'partial work'. In quantitative terms, few highly automated systems exist, around which few highly skilled positions are created. On the aggregate level, the differentiated function structure is thus continued even after the introduction of new technical systems. Improvements occur for parts of the work force, but in total the polarized structure remains. The predictions of the 'new working class' by Mallet and Blauner's 'U-curve', are explicitly rejected (Kern and Schumann, 1970, p. 219 and p. 137).

Some years later, Mickler, Dittrich and Neumann (1976) adopted a similar position in the (German) debate. They studied several plants within three different process industries with differing degrees of automation. One of their main findings was, that given a certain technical system, the division of labor showed little variation because the managerial rationalization strategies are the same (1976, p. 452). Between the three branches different labor organizations were found, whereby the more advanced systems were combined with labor organizations occupying relatively increasing proportions of skilled workers. Although they acknowledge the existence of degrees of freedom, little of this space was used as management aims to reduce labor costs once a specific technical system is installed. Furthermore it was concluded, that all plants were characterized by a polarized structure. Part of the unskilled positions were created as a result of 'mechanization holes', *Mechanisierungslücken*, i.e. residual tasks and work spots which were not included in the design of the technical system.

Huijgen, Riesewijk and Conen (1983, pp. 24-30) refer to many authors in the 1970s, predominantly of German origin, whose empirical work pointed to either degradation or polarization. Much of this empirical work concerns office automation, which were combined with respectively degradation or polarization. Meanwhile, in France Marxist political economists had reserved space for developments with respect to technical systems and work content in theoretical studies about the development of capitalist economies. In the theoretical study *A Theory of Capitalist Regulation*, Aglietta (1976) discusses Fordism and Neo-Fordism. Taylorism is superseded by Fordism, wherein the labor process is characterized by the semi-automatic assembly line c.q. conveyor-belt. Such technical systems led to major labor savings and fixed the workers rigorously to its configuration. Neo-Fordism, a term coined by Palloix, 'is still in an embryonic stage' (1976, p. 122). It is dependent on 'automatic production control', which is more or less used as a synonym for NC, and on the 'recomposition of tasks'. This last development had started in the mid 1960s, which was made possible thanks to improved information systems allowing management to keep control, but 'job rotation and job enrichment are simply the ultimate extensions of Fordism and Taylorism' (1976, p. 128). Aglietta's judgement about automatic production control is slightly different. It eliminates the need for skilled personnel (on the shopfloor), but the engineering department is expanded. The capitalist class has to make sure, that it controls the essential production factors, among which the 'total submission of the highly skilled personnel responsible' for programming, research methods and information processing (1976, p. 127).

Braverman's book *Labor and Monopoly Capital* (1974) evoked a lengthy academic debate on 'the labor process', predominantly in the Anglo-Saxon world. The book was an attempt to demonstrate the overwhelming impact of Taylorism c.q. Scientific Management had had on work content. A multitude of unskilled jobs had been created, enabling managers to control the workforce directly. Braverman's work was later criticized and refined. Friedman (1977) objected, that worker resistance had allowed craft workers to retain part of their autonomy. These craft workers form the core personnel, and are controlled by means of a strategy of 'responsible autonomy', which gains worker loyalty by a mixture of measures, such as higher wages, status, responsibility and job security. The lesser skilled peripheral workers, however, can be controlled by means of direct control. In the course of the debate, a host of other more sophisticated ways of control were described, for instance by Buroway and Edwards (see Christis, 1983 and Thompson, 1989 for an overview). Vital for the discussion on work content is, that the adherents of the Anglo-Saxon labor process approach focused their attention on the politically interesting issue of control rather than on work content itself. Noble's position on the development of the numerical control technology (see section 2.2) fits in this conclusion, as NC was used in his view as a means to transfer the craftsman's production knowledge to management.

Recently, there have been indications that the supposed trend toward an ever increasing division of labor has been stopped and is even reversed. In Germany, Kern and Schumann managed to remain in the very centre of the debate with their second major work, *Das Ende der Arbeitsteilung?* (1984; see Kern and Schumann, 1987, for a summary in English). The bulk of the empirical work was carried out in the automobile, machine tool and chemical industries, all prospering at the time of the investigation. Some branches experiencing worsened business conditions were also included. In the prosperous industries, management more and more came to realize that the borders of Taylorism had been reached and that the elimination of human work in the production process was not the only road to go by, which is presented as a fundamental shift in paradigms: so-called 'New Production Concepts' are emerging. New technical systems are being utilized, which can not (except for machine-feeding jobs) be operated by unskilled workers. Thus, the revised thinking about utilizing humans is connected to the introduction of sophisticated machines. Adaptable workers are needed, who are able to take decisions on their own. These shifts are related to trends on the product markets, namely the demand for an increased variety of quality products, which makes the new production concepts economically viable. Yet, there are also losers, such as the unemployed, and those employed in branches affected by the recession facing possible unemployment. Furthermore, even in the prosperous sectors not all employees are included in the regradation process, although they are likely to tolerate the improvements for their colleagues. These developments lead Kern and Schumann to conclude, that the polarization process has been transformed into a segmentation process, dividing those who benefit from those who do not. Furthermore, they emphasize, that the new production concepts, within the boundaries of segmentation, are likely to set a trend for the future. Follow-up research (Schumann et al., 1990) confirmed that this trend is gaining importance.

A similar argument was developed by the American authors Piore and Sabel (1984), who signalled a 'second industrial divide' if the trend toward 'flexible specialization' is continued. Their message is, that the limits of the industrial developments based on mass production have been reached, and that this dominant strategy is likely to be replaced by a return to craft production. This alternative development path has consequences for the deployment of capital and labor, but also requires the creation of regulatory mechanisms, i.e. the shaping of an institutional environment in which craft production can flourish ('regulation' thus has another meaning than in the French theory of regulation). They argue, that mass production never completely replaced craft production (it persisted in specific regions and industries), although the economic successes of mass production made craft production a discredited paradigm. This may change with the advent of computer controlled machines, which are operated by skilled workers. These new technical systems alone do not suffice to explain the second industrial divide as computerized machinery may be used rigidly, and conventional machines have been used in craft production before. 'If the computer appears to be the cause of industrial flexibility, this is probably less because of its applications than because, malleable as it is, it has helped to crystallize the vision of a flexible economy just as the costs of rigidity were becoming obvious' (Piore and Sabel, 1984, p. 262). However, these new machines can fulfil shifting market demands, which move away from mass production and allow producers to specialize on market segments. Given the response their book generated in sociological literature, it is remarkable, that Piore and Sabel pay relatively little attention to the division of labor. Lane (1989, pp. 163-195) picked up the thread, and provides a more elaborate discussion of flexible specialization in Germany, France and Britain from the perspective of skills. Consistent with the work of Maurice, Sorge and Warner (1980), she concludes, that its wide spread presence of skilled labor gives Germany a head start over the other two countries to compete with a large variety of quality products. With respect to automobile industries, Mueller (1992) concludes, that upgrading of operators' jobs is taking place in many European countries, among which Germany and Britain. In their book on 'lean production', Womack, Jones and Roos (1990) warn against re-installing craft production, in their terms 'neo-craftsmanship', in automobile production because of its inefficiency. On the other hand, one of the advantages of the new 'lean production' above its predecessor mass production is, that it calls for broadly deployable, i.e. higher skilled, employees who take initiatives. Furthermore, the active role of the assembly workers in lean production is seen as more challenging and fulfilling than that of the 'neocraftsmen'.

Although it is beyond the scope of this study to give even a brief overview of the current debates, which were invoked by theses of new production concepts and flexible specialization, some comments on this contemporary literature are useful. In the first place, the authors of both works are extrapolating a trend toward the future, which is a striking similarity with the literature of the 1950s and early 1960s. The main difference between both periods is, that the role of new technical systems as eliminator of unskilled work is now replaced by integrative movements caused by shifting market demands, which can be fulfilled thanks to a combination of new technical systems and skilled labor. For the first time in history, regradation is claimed to be occurring because of a changed design orientation. Secondly, the authors are cautious about their projections and explicitly deal with the limitations. However, much, if not all, of this cautiousness regularly goes lost when their work is discussed in other publications, and the cautious undertone is drowned by the optimistic sound of the tune. This is especially noticeable in the attention for the position of craft labor in the work of Piore and Sabel.

Although, admittedly, the overview of theoretical developments leaves many open ends, some more concrete conclusions can be drawn about the period from 1950 onward.

In a historical perspective, the multitude of differing opinions forms an important constant. In the 1950s and mid 1960s, optimistic visions prevailed, primarily because of the belief that the proceeding automation of production processes would ultimately lead to the elimination of unskilled work. Perhaps in connection with changed economic conditions and\or a changing *Zeitgeist*, i.e. flavored by at that time popular Marxist interpretations, the polarization and degradation tendencies drew most attention in the 1970s. From the 1980s on, the mood changes cautiously to more optimistic views. Gahan sees the emergence of new production concepts as a 'reaction to pessimistic predictions of deskilling and degradation as continuing processes under (monopoly) capitalism, that the optimistic vein which runs through the new production concepts has taken root' (1991, p. 159). In this transition, two related trend breaks were signalled, namely the increasing importance of individualism on product markets and, related to this, the cautious breakthrough of integrative design orientations and methods. Yet, for any period it is virtually impossible to draw a sharp dividing line between the factual developments from the leading thoughts about these developments.

Many linguistically delineated intellectual islands seem to co-exist with a lack of ferries. Roughly, scholars writing in either one of the languages English, French and German have little knowledge of the theories and ideas if published in another than their native language. Exceptions include Blauner's knowledge of Touraine's work, and more generally the impact of translated work or reviews (most notably French work published in Britain), but overall the international exchange of ideas is disappointing, leading to parallel discussions carried on in different linguistic areas. English being the first world language, especially American and British authors have had the chance to dominate the debate as carried on in international literature, thereby largely neglecting the input from other areas. Of course, to a lesser extent the same can be said about this text, not having paid attention to literature from industrialized countries using Spanish, Italian, Japanese or Swedish as their prime language. The obstacle formed by the language is not just a serious barrier for the advancement of knowledge in itself, but also aggravates the already complicated issues in international comparisons. Is there a universal tendency toward regradation, driven by an international individualization of consumer demands as suggested in the 1980s, or do developments in different countries diverge, as suggested by the German emphasis on polarization in the 1970s when the Anglo-Saxon world was more concerned with degradation ?

The academic, professional and political backgrounds of the authors vary, which is likely to influence, consciously or unconsciously, the interpretation of their findings, and the research interests and strategy. This is especially striking in the case of Marxist authors. No matter the period, academic discipline or country of origin, Marxist approaches are not really concerned with work design or technical developments. In a capitalist society, both are subject to capitalist control, which is detrimental to the interests of the working class. This point of view is reflected in Lenin's changing opinion about Taylorism after the Russian revolution (see section 2.1), and repeated time and again. Deskilling of workers enables direct control by management (Braverman), technical systems enable deskilling (Noble), integrative design methods are mere extensions of Taylorism (Aglietta), and if regradation is truly expected, it will lead to a new, revolutionary working class (Mallet). Many of the Marxist publications can be better characterized as lengthy essays, placing 'evidence' in a radical perspective (see section 1.1) rather than empirical work.

Related to this point are the constantly recurring futuristic visions, which generally display an optimistic outlook (Touraine, Kerr, Blauner in the 1950s and 1960s, Bell in the 1970s, and from the 1980s onward the new production concepts and flexible specialization). The optimism about the future contrast sharply with the practice of using the romanticized image of the nineteenth century craftsman as a reference point to demonstrate the overwhelming impact of deskilling. In both cases, the argumentation is not seldom strengthened by appropriate rethorics, creating the impression that it is the phraseology that changes, not the underlying trends. These newly invented terms may rejuvenate the debate, but at the same time, an already acquired body of knowledge is in danger of falling into oblivion.

Despite the more sophisticated research techniques and data, that have gradually become available c.q. been developed, and the widening coverage of work situations, the problem of selectivity in the empirical foundations, as described above for the period until 1911, remains an important issue. The studies containing a smaller or larger number of cases vary with respect to the industries, sometimes even departments, and technical systems covered. Unfortunately, although the authors generally specifically address the limited generalizability of their findings, these limits are seldom spelled out if their studies are successful in the terms of publicity in the academic or journalistic world.

A solution for many of the problems addressed above could at first sight be found in aggregate studies. However, these are faced with a host of other problems. Jaffe and Froomkin (1968) concluded, that little net change occurred in the composition of the American working population, whereas Rumberger (1981) found what could be termed a reversed polarization, i.e. the proportion of middle-level skills grew at the expense of lower- and higher-level skills. Spenner (1983) suggested, that differing research designs yield varying conclusions, whereby case studies show more volatile results than aggregate studies, which generally conclude that 'little-net-change' occurs. A problem in interpreting aggregate studies is, that the trend found, no matter its direction, is the resultant of many determining, sometimes interdependent factors (Lee, 1982; Huijgen, Riesewijk and Conen, 1983). Even when ignoring methodological issues, degradation and regradation can be caused by structural or cyclical changes in employment in industrial branches, shifts in the occupational composition of these branches, and reskilling and deskilling of individual jobs, which on its turn may or may not be connected with the introduction of technical systems. Even methodologically sound and representative quantitative studies fail to identify the importance of technical and organizational changes. Perhaps qualitative and quantitative studies are best used in conjunction, whereby the aggregate studies give a general oversight of trend, which can be explained by means of case studies.

Form et al. (1988), having signalled the indeterminacy of the skilling debate, plea for a contingency approach, whereby these predicted developments on distinguished levels are dependent on specified factors. In the following, the level of the individual, in this case robotized workspot is further analyzed in the perspective of degradation and regradation.

The robotized workspot

In chapter five, the output characteristics were assigned a prominent role, determining the frequency with which specifiable machine-related tasks occur. They delineate the quantity of a substantial part of the tasks that have to be performed to realize the output. Yet, the coupling between output characteristics and work design was not so close, that dynamic output characteristics automatically led to integrated functions. Rather, such a situation offers the possibility to either decentralize or centralize programming. Both choices were found to be effectuated, which is a matter of organizational design orientations. The evidence suggests, that the output characteristics and the organizational design orientation are among the most important contingency factors influencing job design around arc welding robots. One wonders what exact role the output characteristics play around other machines, such as CNC-machines and FMSs, and at higher levels of aggregation. The empirical evidence in this respect is indicative. Figure 4.1 makes it credible, that these features play a role in the decision to (de-)centralize programming in case of CNC-machines, and low programming frequencies were repeatedly mentioned as 'adding little to job content', but the impact of the output characteristics has not been systematically investigated in case of CNC-machines c.q. FMSs, with the exception of batch size.

Some additional literature may serve to place the empirical findings of chapter five in the perspective of the literature discussed above and to nuance the robot-friendly image commonly found in technical literature. To start with the latter, the use of robots is often promoted to improve the quality of working life, especially with respect to work content and working conditions. Two quotations in the 'Industrial Robot' discussing robot statistics illustrate this point:

> The main objective of an industrial robot is to free human beings from repetitive tasks

and

> Human operators will move between supervision of low-level automation to direct control of complex operations (Industrial Robot, 1991).

The latter quotation will sound familiar having dealt with the ideas of Touraine and Blauner. Although it is perhaps used unconsciously, the use of 'will', referring to future events, forms a striking similarity with the future-oriented optimism, which was found to occur repeatedly. As is argued in the following, both quotations need balancing. Subsequently, work content and working conditions are discussed.

Work content often changes favorably after robots have been introduced. Monotonous handling and assembly work can also be performed by robots (Ebel, 1986; Okubayashi, 1986, p. 30). On the other hand, negative conditions are frequently mentioned in combination with robots. These concern the formation of monotonous, repetitive, machine-paced, unskilled, and isolated work (Ewersheim, Bette and Hausmann, 1986, p. 482; Mossink and Peeters, 1992). The monotony is sometimes worsened due to increase machine pacing and more intense levels of working (Fleck, 1983, p. 50).

Yokomizo makes the following statement about the skills needed to operate automated equipment, among which robots:

> Middle-aged, older and mentally handicapped workers who are not suited for other jobs are also engaged in simple jobs. For example, the mentally handicapped run numerically controlled machines and machining centres, older workers and those not fit for other tasks are engaged in inspection and loading parts onto robotic machines, middle-aged and older people work on conveyor lines, including automated machinery (1986, p. 170).

Whereas this quotation may be exaggerated and was probably written with the Japanese situation in mind, the example of Petra (see sections 5.2 and 5.3) proves that 'less valid' persons suffice to conduct operating tasks. Negative consequences on job content and subsequently skill levels are especially likely to arise around stand-alone robots, of which arc welding robots are the most widely occurring example. If not all operating tasks are automated, the bulk of the work around stand-alone robots is likely to be simple (Wobbe, 1989, p. 3). In this respect, two German studies are interesting, namely the ones by respectively Lauenstein, Seitz and Volkholz (1989) and Urban and Hölldampf (1988).

Lauenstein, Seitz and Volkholz (1989) analyzed thirteen cases, where robots were used for a variety of applications (five cases concern arc welding). These thirteen cases are grouped in three types (respectively represented by three, six and four cases), for which purpose four characteristics (length of the cycle time, the degree of the division of labor, the degree of routine work, and the changes after the introduction of the robot) are used. In Types I and II, the working situations have undergone no basic changes before and after the introduction of the robot (see Table 6.1).

Table 6.1
Types I and II of Lauenstein, Seitz and Volkholz

Type	**I**	**II**
Cycle time	< 90 sec.	> 2 min.-2.5 hrs.
Division of labor	high	lower than in Type I
Percentage of routine activities	70 - 90	50 - > 70
Changes	no	no

Source: Compiled from Lauenstein, Seitz and Volkholz, 1989, pp 61-86

In both Types I and II, the formerly existing function structure has more or less been replicated in the new situation, with Type I clearly being less attractive (the cases with stable output characteristics are representatives of Type I: monotonous manual welding has been replaced by even more monotonous (un-)loading). In the cases of the Type II, several work organizations are possible: if the workload is sufficiently high, the simple tasks can be grouped into an unskilled function, but there are also possibilities for team work.

The function structure has changed in the cases of Type III. The initial situation was similar to the one of Type I, namely much routine work, short cycles, few skills demanded, and a high division of labor. The main difference between both types is, that in Type III the majority of routine and machine paced work was taken over by the robotic system, leaving system monitoring as an important tasks for operators. A precondition for the implementation of Type III robot systems is, that the work is routine and short cyclic. Otherwise, it is technically difficult and thus expensive to robotize the production process. Programming generally only occurs when the system is implemented. Resolving disturbances and optimizing the system are important tasks during the initial periods of operation. The cognitive skills needed for monitoring are sometimes even below the level needed for the manual work in Type II. Monitoring can either be a residual task or part of the job of a skilled system operator.

In all cases, improvement measures can be taken, although it is noteworthy, that only two measures affecting work content are mentioned, namely the use of buffers and task rotation outside the robotic system.

With respect to the creation of integrated functions in two out of four welding cases of Type II, Lauenstein c.s. note, that it is necessary that programming takes a substantial amount of time.

Urban and Hölldampf (1988) concentrate on arc welding robots. They made a classification of arc welding situations, which is based on three factors which influence the type of work:

1. batch size.
2. number of parts to be welded
3. product size.

The combination of these three factors lead to eight different types of production situations. Six of these are presented in Table 6.2. The two remaining types concern combinations of large batches and many parts. They can be broken down in a combination of the cases included in Table 6.2, and hence add nothing to the essence of the classification.

Table 6.2
Classification of robotic arc welding situations

Batch size	Small				Large	
Number of parts	Small		Large		Small	
Product size	Small	Large	Small	Large	Small	Large
Type	1	2	3	4	5	6

Source: Adapted from Urban and Hölldampf, 1988, p 6

If no counter measures are taken, the following disadvantages result as the consequence of these three factors. Large batches lead to very monotonous production situation. Only when small batches are combined with a large number of parts (types 3 and 4), and the division of labor is not high, the authors judge the situation as 'moderately monotonous' (1988, p. 9). A small number of parts means monotonous work, but reduces the risk, that accidents happen during disturbances. Small products, and thus short welding times, lead to machine paced work, generally at isolated work spots. Isolation is 'intermediate', if cycle times are long.

There are technical and organizational (here meant to include personnel policy) measures to improve these situations. Technical measures include the use of buffers to decouple the operator from the robot's work rhythm and lengthening the cycle time, although Urban and Hölldampf do not specifically mention the possibility to combine several products in one production run nor the combination of more than one product in one fixture (see the case of Component in section 5.3). Furthermore, the use of lifting devices would reduce the problem of handling sizeable products (and\or fixtures). The organizational measures basically come down to job rotation, whereby the burden of the monotonous work is shared by several operators, who are preferably members of a production team. Job design is mentioned as a more general strategy, but does not seem to be linked with any particular production situation.

Although their classification does not cover all output characteristics (especially product variability is a noteworthy omission), the introduction of arc welding robots may have a favorable impact on the working situation, for instance by increasing the distance from human being to the welding process, reducing the risk of inhaling noxious fumes. Furthermore, welding large batches of simple products is not likely to be considerable pleasurable, whether the process is done manually or by a robot. Nevertheless, whether the introduction of robots has lead to improved working conditions can only be assessed by comparing the former and the current situation. The new robotized situation is likely to entail unfavorable circumstances, unless explicit attention has been paid to avoid these (Urban and Hölldampf, 1988, pp. 8-9). This conclusion is similar to the general conclusion with respect to job design: unless the organization follows some form of integrative design principles, fractionized jobs are likely to occur.

In conclusion, the introduction of robots does not necessarily lead to an improvement of work content. In the cases of Lauenstein c.s., the former work organization basically remains unchanged after the introduction of robots unless the production process is completely robotized. In the latter case, monitoring the robot system becomes the main system related task. As arc welding robots are without exception fed manually, much unskilled work remains. In these cases, the output characteristics determine to an important degree the frequencies with which the robot-related tasks occur (see Types I and II of Lauenstein c.s., all cases of chapter five and those by Urban and Hölldampf). Where stable output characteristics exist, robot operators are likely to perform monotonous and unskilled work. Where dynamic output characteristics predominate, different organizational options are chosen. This situation is a necessary, but insufficient precondition for the creation of integrated functions. Thus, the conclusion can be drawn, that it is not so much the industrial robot that determines whether or not work content is improved, but more the application and production use of this robot in combination with organizational design. Unless explicit attention is paid to the improvement of work design, unfavorable results are likely to occur, a finding which also holds for working conditions.

Situations with bad **working conditions** are often improved by having robots take over taking over dirty, dangerous or heavy work (Ebel, 1986, p. 47). Mossink and Peeters (1992) mention the example of a robot which (un-)loads furnaces, handling hot workpieces. Forging, spray painting, foundry work, and machine tool loading are mentioned by Lambrinos and Johnson (1984, p. 24) as other examples where robots have removed humans from hazardous work environments. Field robots form a relatively new development in this area, serving as bomb detectors or as cleaners of nuclear waste. Yet, some related critical remarks need to be made in this respect. In the first place, these applications typically represent a relatively small proportion of robot applications. The two foremost important applications concern spot and arc welding. For welding, the favorable impact of using robots is merely the increased distance between weld and human, lowering dangers due to the emission of radiation from the arc and inhalation of possibly noxious fumes (Mangold, 1991, p. 72). Yet, frequently insufficient safety precautions are taken to protect operators from dangerous situations (Mossink and Peeters, 1992), occasionally resulting in deadly accidents (Nagamachi, 1986, p. 341). Unless attention is paid to safety aspects, robotized work spots may be just as well be connected with dangerous than improved working conditions (cfr. Ebel, 1992, pp. 106-107). Secondly, such bad working conditions are probably connected to 'labor problems', such as high absenteeism, high turnover, and recruitment problems (Fleck, 1983, p. 50). Labor shortages are with a certain regularity mentioned as reason to purchase robots, and the leading position of Japan as robot user is partially ascribed to its labor shortage. In the third place, the 'humanization effect' is sometimes just an accompanying effect, as Mickler c.s. (1981, p. 268) concluded from their research in the German Volkswagen plant at Wolfsburg. In all cases, the use of robots was economically justified, but their contribution to the humanization of work has to be considered a welcome side-effect. No robots were implemented without the fulfilment of economic conditions. Yet, the favorable impact on humanization has lead to a positive attitude toward robots. This last finding is confirmed by Okubayashi for Japan: 'The initial limitation of robots to taking over the 'dirty' work in industry resulted in their acceptance by the workers' (1986, p. 24). Resuming, unfavorable working conditions can indeed be improved by using robots, but this motive only concerns particular applications and has to be seen in connection with economic and labor problems. Elsewhere, repetitive tasks and functions come into existence after robots have been introduced, and safety measures are not always adequate.

Approximately five years later than Mickler c.s. (1981), Landau and Brandenburg visited Volkswagen's Wolfsburg plant again. Their final conclusion is still valid:

> The use of industrial robots does not necessarily lead to improved working situations. Both positive and negative effects occur. There are possibilities for improvement, but their realization depends on managerial goals and paradigms (1986, p. 204).

The problems encountered in the early research can still be found. The negative consequences have not gone unnoticed, and measures have been developed to eliminate or alleviate these problems. Urban and Hölldampf make may useful suggestions to ameliorate robotized arc welding, and many of these were included in a brochure from the German union *IG Metall* (1986). The same can be said of Mangold's (1991) recommendations to improve safety conditions. The Dutch Labor Inspectorate has issued a brochure for this purpose (see *Industriële robots*, 1991). However, integrated functions are far from being the standard, and the safety conditions in many organizations leave much to be desired. The question arises why the efforts to improve working conditions seem to have little effect, but the answers remain speculative. Many organizations may be deterred by the improvement costs, the publications may not reach the target audience, a technical preoccupation may still predominate, or many organizations do not really see alternatives to produce large batches of simple products (to name just one production situation which is likely to lead to an unfavorable working situation).

Back to the **skilling debate**. The introduction of the technical system 'industrial robot' is generally a replacement of a manual welder by a robot-man system. Therefore, the conclusion that the output characteristics and the design orientation play an important role in the formation of work around robots is only partial, as the initial situation has so far not been addressed. However, this is necessary to draw conclusions with respect to deskilling or reskilling.

The role of the output characteristics in the initial situation is important because of two reasons. In the first place, the production of the manual welder can also be characterized by output characteristics. Constantly welding long batches of simple products can not be considered to be attractive work. When a robot-man system is introduced to replace such a situation of stable output characteristics, the main change in the function structure is the replacement of a semi-skilled welder by an unskilled robot operator: degradation is the result. In fact, this change is an example of the classical substitution from manual work to repetitive feeding work (Kern and Schumann, 1970). In the second place, robotized welding is economically only feasible for a part of the production, namely for those products which are technically suited to be welded by a robot and which are produced in sufficient quantities. In this case, the robot takes over routine production, relieving manual welders from monotonous work. The welders may experience a slight increase in welding skills needed, especially those who get to program the robot, but the main effect is that unskilled robot operators are needed to feed the robot, i.e. degradation occurs. Only when fully automated systems, leaving no residual tasks, are created, regradation occurs at the level of the organization because repetitive semi-skilled work is eliminated or replaced by monitoring, taking little time and generally added to an existing job. Given technical difficulties and financial barriers, full mechanization is only likely to occur when large numbers of preferably almost identical products are produced, i.e. stable output characteristics.

Only where integrated functions are created, thus in a situation of dynamic output characteristics where an integrative design was chosen, regradation in the sense of reskilling occurs. In this case, a skilled welder has to increase his craftsmanship by learning the robotized production process.

Table 6.3 summarizes the effects of the introduction of industrial robots in the situations distinguished on the function structure. It concentrates on the operator's function as the robot-related tasks of others generally make up only a fraction of their total workload (Haring and de Mare, 1991).

Table 6.3
Industrial robots and function structure

Output characteristics in new situation	Jobs in new situation	Change in function structure
Stable	No personnel left	(Regradation)
	Only operators	Degradation
Dynamic	Separate programmer	Polarization
	Operator-programmer	Regradation

Note : In case of 'no personnel left', the parentheses around regradation point to the fact, that regradation occurs at the departmental level, not at the level of the workspot.

Table 6.3 is basically similar to Table 5.4, which links the output characteristics to the organizational options which were found to be chosen in the case-studies. The integrated function of Table 5.4 is called 'operator-programmer' in Table 6.3, and following Lauenstein et al., an additional possibility is included in Table 6.3, namely 'no personnel left'. As said, this option was not found around arc welding robots, but may occur around other applications.

It follows from Table 6.3, that the robot itself is less important for (changes in) the function structure than the situation in which the robot is placed and the organizational design orientation adopted. This finding confirms hypothesis 1, that the technical system and the organizational design orientation are independent. Hypothesis 2, which links the output characteristics to the occurrence of integrative organizations is partially confirmed and nuanced. The focus here is at the workspot, not at the level of the organization as in hypothesis 2. However, drawing a parallel between both levels of aggregation, a link between output characteristics and organizational options is present: only dynamic output characteristics provide the option to choose for integrated functions. Thus, the existence of a relationship between both factors is confirmed. However, the character of this relationship is different from the one implicitly assumed in hypothesis 2. Where this hypothesis implicitly departs from the **realization** of the option to create integrated functions, the evidence merely shows the **presence** of the option. Its realization is a matter of adopting an integrative design orientation. Thus, the design orientation provides another point to choose. By stabilizing the output characteristics, an ideal situation may be created for either eliminating unskilled or creating such work. This last option forms the quintessence of hypothesis 3, but was not found to occur (see section 5.3). However, as only part of an organization's total product range is selected to be welded manually, an analogous development may take place. The output characteristics of robotized products are likely to be more stable than those of the remaining, non-robotized products.

It can be concluded, that robotized work situations do not fit either of the categories regradation, degradation or polarization. Instead, the specific outcome is to an important extent dependent on contingency factors such as the output characteristics and the design orientation. The only empirical affirmation of a trend toward regradation was found in the three cases of chapter five with integrated functions. However, in the majority of cases unskilled work forms the major part of robot-related work. Furthermore, several of the factors affecting the function structure that were mentioned in the review of literature, still have not lost their relevance. As far as industrial robots are concerned, the notion defended by Levasseur, Marshall and later several authors in the 1950s, that technical systems eliminate unskilled work, is found to be just as true as the opinion of their opponents (Sismondi; Marx; Kern and Schumann, 1970), that unskilled feeding work is created.

These findings raise the question whether output characteristics have been given sufficient attention in earlier research. Admittedly, the rise of mass production in the nineteenth century and recently its decline have been associated with changes in work content, but this work directs its attention to the organizational level. The same holds for Burns and Stalker (1961). However, studies on the introduction of new technical systems seem to underestimate the role output characteristics play. Batch sizes practically form the only exception to this rule (see chapter four). If these output characteristics are given due attention in the future, it is important to study them at the appropriate level. There are few problems to be expected if this level is the workspot, but at higher levels of aggregation difficulties may arise as decoupling between the output characteristics at the differing levels may occur. Thus, even within one organization two identical machines can be used in totally different ways. One might have expected intuitively, that a flexible machine as the industrial robot would fit well in an era where mass production is said to lose its dominance and a return to craft production is envisioned. If robots are seen as flexible machines, it has to be realized, that an inherent characteristic of flexibility is its optional value. The flexibility may be used, but the execution of this option is by no means a necessity. This notion may be counter-intuitive when applied to a technical system with an aura of multi-functionality such as the robot. Unfortunately, it is apparently less hard to comprehend when applied to the even more flexible human being.

Bibliography

Adler, P.S. (1988), 'Managing Flexible Automation', *California Management Review*, 30(3), pp. 34-50.

Adler, P.S. and B. Borys (1989), 'Automation and Skill: Three Generations of Research on the NC Case',*Politics and Society*, 17(3), pp. 377-402.

Aglietta, M. (1976\1979), *A Theory of Capitalist Regulation; The US Experience*, New Left Books, London.

Alders, B. and J. Christis (1988a), 'Arbeid en recrutering in de machinebouw of 'Hoe efficiënt wordt de CNC-technologie in de praktijk toegepast ?'', *Tijdschrift voor Arbeidsvraagstukken*, 4(4), pp. 72-87.

Alders, B. and J. Christis (1988b), 'Technologie, arbeidsorganisatie, en arbeidsmarkt', *Tijdschrift voor Politieke Ekonomie*, 11(2), pp. 97-121.

Alders, B., J. Christis and R. Bilderbeek (1988), *Technologische ontwikkeling en veranderingen in de werkgelegenheidsstructuur; Samenvattend rapport*, Ministry of Social Affairs and Employment, The Hague.

Alders, B., B. Fruytier and R. Peters (1991), *Sociotechniek en de drempel van de technologie; De mogelijkheden en onmogelijkheden van CAD- en PPS-systemen*, in: POST-groep, pp. 69-85.

Aldrich, H. (1972), 'Technology and Organizational Structure; A Reexamination of the Findings of the Aston Group', *Administrative Science Quarterly*, 17(1), pp. 26-43.

Allegro, J.T. (1973), *Socio-technische organisatieontwikkeling*, Stenfert Kroese, Leiden.

Allegro, J.T. and H.W. Kanters (1975), 'Een experiment betreffende het herstruktureren van taken in een gesimuleerde organisatie-setting', *M&O*, 29(4), pp. 189-208.

van Amelsvoort, P. (1985), 'Arbeidsorganisatie en computer bestuurde bewerkingsmachines', *Technische Bedrijfsvoering*, 19(6), pp. 172-177.

ANERTEK (1989), *Robots in Nederland; De robotsituatie per 31-12-1988*, ANERTEK, Bilthoven.

Argyris, C. (1957), *Personality and Organization; The Conflict Between System and the Individual*, Harper & Brothers, New York.

van Assen, A. and J.F. den Hertog (1980), *Werkbeleving en werkstrukturering*, in: de Galan, van Gils andvan Strien (eds.), pp. 49-77.

Astley, W.G. and A.H. Van de Ven (1983), 'Central Perspectives and Debates in Organizational Theory',*Administrative Science Quarterly*, 28(2), pp. 245-273.

Ayres, R.U., R. Dobrinsky, W. Haywood, K. Uno and E. Zuscovitch (eds.) (1992), *Computer Integrated Manufacturing; Volume IV: Economic and Social Impacts*, Chapman & Hall, London\New York\ Tokyo\Melbourne\Madras.

Ayres, R.U., W. Haywood and I. Tchijov (eds.) (1992), *Computer Integrated Manufacturing; Volume III: Models, Case Studies, and Forecasts of Diffusion*, Chapman & Hall, London\New York\Tokyo\ Melbourne\Madras.

Babbage, C. (1835\1971), *On the Economy of Machinery and Manufactures* (fourth enlarged edition), Augustus M. Kelley Publishers, New York.

Babüroglu, O.N. (1988), 'The Vortical Environment: The Fifth in the Emery and Trist Levels of Organization Environments', *Human Relations*, 41(3), pp. 181-210.

Bainbridge, L. (1983), 'Ironies of Automation', *Automatica*, 19(6), pp. 775-779.

Batenburg, R.S. (1991), *Automatisering in bedrijf; Een empirisch-theoretisch onderzoek naar de effecten van automatisering op de functiestructuur van bedrijven*, Thesis Publishers, Amsterdam.

van Beek, H.G. (1964), 'The Influence of Assembly Line Organization on Output, Quality and Morale', *Occupational Psychology*, 38(4-5), pp. 161-172.

Beer, M., B. Spector, P.R. Lawrence, D. Quinn Mills and R.E. Walton (1984), *Managing Human Assets; The Groundbreaking Harvard Business School Program*, The Free Press, New York.

van Beinum, H.J.J. (1963), *Een organisatie in beweging; Een sociaal-psychologisch veldexperiment bij de postcheque- en girodienst*, Stenfert Kroese, Leiden.

van Beinum, H.J.J., M.R. van Gils and E.J. Verhagen (1968), *Taakontwerp en werkorganisatie; Een socio-technisch veldexperiment*, NIPG-TNO, Leiden.

Bell, D. (1973), *The Coming of Post-Industrial Society*, Basic Books, New York.

Bell, J.H.J. (1990), *Profiteren van de profits ?*, unpublished Master's thesis Tilburg University, s.n., Tilburg.

Benders, J. (1990), 'Over de inzetbaarheid van werknemers', *Tijdschrift voor Politieke Ekonomie*, 13(2),pp. 94-106.

Benders, J. (1991a), 'Het gebruik van industriële robots', *Informatie*, 33(11), pp. 798-807.

Benders, J. (1991b), *Dividing Labour around Robots; A Theoretical Model*, unpublished paper for the WESWA-conference, Tilburg University, Tilburg.

Benders, J., (1992), *Towards a Temporal Division of Labour ?*, in: Hollier, Boaden and New (eds.), pp. 149-154.

Benders, J. and J. de Haan (1991), *The Division of Labour around Industrial Robots*, in: Karlsson and Rubenowitz (eds.), supplement: pp. 1-18.

Benders, J. and J. de Haan (1992), *Robots and People: A Theoretical Model*, in: Khalil and Bayraktar (eds.), pp. 817-827.

Bengtsson, L. (1992), 'Work organization and occupational development in CIM: the case of Swedish NCmachine shops', *New Technology, Work and Employment*, 7(1), pp. 29-43.

Bennett, D.J. (1992), *Manufacturing Flexibility, Technology and Organization*, in: Khalil and Bayraktar (eds.), pp. 777-786.

Berenschot (1989), *Industriële robots; FMS- en FMC-systemen; Huidige situatie, toekomstperspectief en beleid voor Nederland*, Ministry of Economic Affairs, s.l. [the Hague].

van Berkel, P. (1968), *Beleidsinvloed op organisatie en arbeidsvoldoening*, in: Berting and de Sitter (eds.), pp. 97-125.

Berting, J. and L.U. de Sitter (eds.) (1968), *Arbeidsvoldoening en arbeidsbeleid*, Marka, Antwerp.

Bilderbeek, R. and B. Alders (1988), *Advanced production systems and training implications*, in: Buitelaar(ed.), pp. 182-194.

Bilderbeek, R.H. and P.J. Kalff (with cooperation of F. Prakke) (1985), *Automatisering van de fabricage; Een verkenning van technische, economische en sociaal-organisatorische aspecten*, Kluwer, Deventer.

Blauner, R. (1964), *Alienation & Freedom; The Factory Worker and His Industry*, University of Chicago Press, Chicago\London.

Bloemen, E.S.A. (1988), *Scientific Management in Nederland 1900-1930*, NEHA, Amsterdam.

Boekraad, H., G. van den Brink, H. Hoeks and F. van Wel (eds.) (1983), *Te Elfder Ure 33*, 27(1).

Boekraad, H.C., G. van den Brink and F. van Wel (eds.) (1988), *Te Elfder Ure 41*, 29(3).

Bolwijn, P.T., J. Boorsma, Q.H. van Breukelen, S. Brinkman and T. Kumpe (1986), *Flexible Manufacturing; Integrating Technological and Social Innovation*, Elsevier, Amsterdam\Oxford\New York\Tokyo.

Bolwijn, P.T. and T. Kumpe (1990), 'Manufacturing in the 1990s - Productivity, Flexibility and Innovation', *Long Range Planning*, 23(4), pp. 44-57.

Braczyk, H.-J. (1987), *Human Centred Technology, Strategic Options and User's Consent*, in: Brödner (ed.), pp. 143-154.

Braverman, H. (1974), *Labor and Monopoly Capital; The Degradation of Work in the Twentieth Century*, Monthly Review Press, Chicago.

Bright, J.R. (1958), *Automation and Management*, Harvard University Press, Boston.

Brödner, P. (ed.) (1987), *Strategic Options for "New Production Systems" - CHIM: Computer and Human Integrated Manufacturing*, FAST-paper 150, Commission of the European Communities, Brussels.

Brödner, P. (1991), 'Design of Work and Technology in Manufacturing', *International Journal of Human Factors in Manufacturing*, 1(1), pp. 1-16.

Buitelaar, W. (1988), *Technology and Work; Labour Studies in England, Germany and the Netherlands*, Avebury: Aldershot\Brookfield USA\Hong Kong\Singapore\Sydney.

Buitelaar, W. (1989), *The Dynamics in the Relationship between research institutions and trade unions. A report on Belgium, Austria, Norway and the Netherlands*, in: van Tulder (ed.), pp. 67-82.

Buitelaar, W. (1991), 'Over de moderniteit van de vooroorlogse Nederlandse arbeidssociologie', *Tijdschrift voor Politieke Ekonomie*, 14(2), pp. 56-64.

Buitendam, A., D.A.G. Dumas and A.C. Glebbeek (eds.) (1990), *Het Zweedse model: geschikt voor import ?*, Van Gorcum, Assen\Maastricht.

Burns, T. and G.M. Stalker (1961), *The Management of Innovation*, Tavistock,

London.

Burrell, G. and G. Morgan (1979), *Sociological Paradigms and Organisational Analysis*, Heinemann, London.

Butera, F. (1975), *Environmental Factors and Job and Organization Design: The Case of Olivetti*, in: Davis and Cherns (eds.), pp. 166-200.

Buyse, J.J. (1987), *Kwaliteit van werk en organisatie op de produktievloer; Een integrale studie op mikro-nivo*, unpublished PhD-dissertation University of Nijmegen, s.n. [Dorag], s.l. [Beuningen].

Campbell, D.T. and J.C. Stanley (1963 (1966), *Experimental and Quasi-Experimental Designs for Research*, Rand McNally, Chicago.

CBS (1987), *Automatiseringsstatistieken; Industriële automatisering 1984 - 1986*, Staatsuitgeverij, The Hague.

CBS (1988), *Maandstatistiek voor de industrie*, 30(5), Staatsuitgeverij, The Hague.

CBS (1990), *Maandstatistiek voor de industrie*, 32(4), Staatsuitgeverij, The Hague.

CBS (1991), *Automatiseringsstatistieken; Ontwerp- en produktieautomatisering (CAD/CAM) 1988-1990*, SDU, The Hague.

CEDEFOP (1984), *Beschrijving van de beroepsopleidingssystemen in de Lid-Staten van de Europese Gemeenschap; Vergelijkende Studie*, CEDEFOP, Berlijn.

Child, J. (1972), 'Organizational Structure, Environment and Performance: The Role of Strategic Choice',*Sociology*, 6(1), pp. 1-22.

Child, J. (1974), 'What determines Organization?', *Organization Dynamics*, 4, pp. 2-18.

Christis, J. (1983), *Het Braverman debat*, in: Boekraad, van den Brink, Hoeks and van Wel (eds.), pp. 15-58.

COB\SER (1984), *Flexibele automatisering; Kansen of beter werk*, COB\SER, The Hague.

COB\SER (1988), *CNC-machinebedieners, Produktie-organisatie en Flexibele automatisering; Een onderzoek naar de invoering van CNC-machines in bedrijven op flexibiliteit, produktiviteit en produktkwaliteit en de inhoud van het werk*, COB\SER, The Hague.

COB\SER (1990), *Zelfstandig samenwerken in autonome taakgroepen; Praktijkervaringen in industrie en dienstverlening*, COB\SER, The Hague.

Corbett, M. (1987), *Strategic Options for Human-Machine Interface Design*, in: Brödner (ed.), pp. 157-173.

Dankbaar, B. (1988), 'New Production Concepts, Management Strategies and the Quality of Work', *Work, Employment & Society*, 2(1), pp. 25-50.

Dankbaar, B. (1990), *Prospects for Anthropocentric Production Systems in the Netherlands*, FAST-paper 256, Commission of the European Communities, Brussels.

Davis, L.E. (1976), *Job Design: Historical Overview*, in: Davis and Taylor (eds.), pp. 29-35.

Davis, L.E., R.R. Canter and J. Hoffman (1955), 'Current Job Design Criteria', *Journal of Industrial Engineering*, 6(2), pp. 5-11.

Davis, L.E. and A.B. Cherns (1975), *The Quality of Working Life; Volume Two. Cases and Commentary*, Free Press\Collier Macmillan, New York\London.

Davis, L.E. and J.C. Taylor (1976), *Technology, Organization and Job Structure*, in:

Dubin (ed.), 379-419.

Davis, L.E. and J.C. Taylor (eds.) (1979), *Design of Jobs* (second edition), Goodyear Publishing Company, Santa Monica.

De Meyer, A. and K. Bonheure (1991), *Testing the Sand Cone Model Across Industries; A Comparative Study*, in: Karlsson and Rubenowitz (eds.), pp. 183-197.

De Meyer, A. and K. Ferdows (1991), *Removing the Barriers for Manufacturing; Report on the 1990 European Manufacturing Futures Survey*, in: Karlsson and Rubenowitz (eds.), pp. 198-212.

van Dijck, J.J.J., J.L.J.M. van der Does de Willebois and M.J.M. Daniëls (eds.) (1970), *Organiseren rondom mensen (Technologie, economie, autonomie)*, Stenfert Kroese, Leiden.

Dirne, C.W.G.M. (1989), *Planning Problems of Flexible Automated Cells in a Job Shop - A Case Study*, in: Stecke and Suri (eds.), pp. 61-66.

Dirne, C.W.G.M. (1990), *Production Control for Flexible Automated Manufacturing Stations in Low Volume Component Manufacturing*, unpublished PhD-dissertation Eindhoven University of Technology, s.n. [Wibro], s.l. [Helmond].

Docter, H.J. and C.T.M. Stokman (1987), *Diffusie van innovaties: met kennis meer kans; Een onderzoek naar de samenhang tussen informatie en innovatie in het midden- en kleinbedrijf*, EIM, Zoetermeer.

Dodgson, M. (1987), 'Small firms, advanced manufacturing technology and flexibility', *Journal of General Management*, 12(3), pp. 58-75.

Doorewaard, H. and F. Huijgen (1985), 'Van vakspecialist tot all-rounder; Verandering in functie- en opleidingsstructuur in het verzekeringswezen', *Tijdschrift voor Arbeidsvraagstukken*, 1(2), pp. 53-66.

Drucker, P.F. (1954), *The Practice of Management*, Harper & Brother Publishers, New York.

Dubin, R. (ed.) (1976), *Handbook of Work, Organization, and Society*, Rand McNally, Chicago.

Düll, K. (1975), *Industriesoziologie in Frankreich*, Europäische Verlagsanstalt, Frankfurt am Main.

Ebel, K.-H. (1986), 'The impact of industrial robots on the world of work', *International Labour Review*, 125(1), pp. 39-51.

Ebel, K.-H. (1992), *Management, Jobs, and Employment*, in: Ayres, Dobrinsky, Haywood, Uno and Zuscovitch (eds.), pp. 93-120.

Economic Commission for Europe (1986), *Recent trends in flexible manufacturing*, United Nations, New York.

Eichener, V. (1990), *Organisational Concepts in German Industry*, FAST Occasional Paper 270, Commission of the European Communities, Brussels.

van Eijnatten, F.M. (1985), *STTA; Naar een nieuw werkstruktureringsparadigma*, unpublished PhD-dissertation University of Nijmegen, s.n. [Nederlandse Philips Bedrijven], s.l. [Eindhoven].

van Eijnatten, F.M. (1991), *From Autonomous Work Groups to Democratic Dialogue and Integral Organizational Renewal: 40 Years of Development and Expansion of the Socio-technical Design Paradigm*, Research Memorandum 91-015, MERIT, Maastricht.

Eisenhardt, K.E. (1989), 'Building Theories from Case Study Research', *Academy of*

Management Review,14(4), pp. 532-550.

Emery, F.E. and E.L. Trist (1965), 'The Causal Texture of Organizational Environments', *Human Relations*, 18(1), pp. 21-32.

Engels, F., (1845\1952), *The Condition of the Working-Class in England in 1844*, George Allen & Unwin, London.

Ewers, H.-J., C. Becker and M. Fritsch (1990), *Wirkungen des Einsatzes Computergestützter Techniken in Industriebetrieben*, Walter de Gruyter, Berlin\New York.

Ewersheim, W., B. Bette and A. Hausmann (1986), 'Industrierobotereinsatz in der Produktion; Ziele Chancen und Risiken', *Die Betriebswirtschaft*, 46(4), pp. 473-485.

Eyada, O.K. and M.M. Ahmad (eds.) (1992), *Flexible Automation and Information Management*, CRC Press, Boca Raton\Ann Arbor\London\Tokyo.

Farkas, G. and P. England (eds.) (1988), *Industries, Firms, and Jobs: Sociological and Economic Approaches*, Plenum Press, New York.

Fayol, F., (1916\1920), *Administration industrielle et générale*, Dunod, Paris.

Fleck, J. (1983), *Robots in manufacturing organisations*, in: Winch (ed.), pp. 44-56.

FME (1989a) *Verslag van een onderzoek naar scholings- en arbeidsmarktaspecten in de regio Oost-Nederland*, FME, Zoetermeer.

FME (1989b) *Verslag van een onderzoek naar scholings- en arbeidsmarktaspecten in de regio Zuid-Nederland*, FME, Zoetermeer.

FME (1989c, *Verslag van een onderzoek naar scholings- en arbeidsmarktaspecten in de regio Noord-Holland*, FME, Zoetermeer.

FME (1989d, *Verslag van een onderzoek naar scholings- en arbeidsmarktaspecten in de regio Rijndelta*, FME, Zoetermeer.

FME (1990a) *Verslag van een onderzoek naar scholings- en arbeidsmarktaspecten in de regio Midden\West-Nederland*, FME, Zoetermeer.

FME (1990b) *Verslag van een onderzoek naar scholings- en arbeidsmarktaspecten in de regio Noord-Nederland*, FME, Zoetermeer.

FME (1990c, *Verslag van een landelijk onderzoek naar scholings- en arbeidsmarktaspecten*, FME, Zoetermeer.

FME (1992), *Arbeidsmarktonderzoek metalelektro-industrie Zuidwest-Nederland 1992*, FME, Zoetermeer.

Fontijne, J. (1970), *Evolutie, experiment en evaluatie in de produktie-organisatie van een televisiefabriek*, in: van Dijck, van der Does de Willebois and Daniëls (eds.), pp. 32-42.

Form, W., R.L. Kaufman, T.L. Parcel and M. Wallace (1988), *The Impact of Technology on Work Organization and Work Outcomes; A Conceptual Framework and Research Agenda*, in: Farkas and England (eds.), pp. 303-328.

Friedman, A.L. (1977), *Industry and Labour; Class Struggle at Work and Monopoly Capitalism*, MacMillan, London.

Friedmann, G. (1946\1952), *Der Mensch in der mechanisierten Produktion*, Bund-Verlag: Köln.

Fruytier, B. (1988), 'Taylorisme of het nieuwe produktiekoncept. De ekonomische effekten van de invoering van CNC-machines', *Tijdschrift voor Politieke Ekonomie*, 11(4), pp. 79-103.

Fruytier, B. and K. ten Have (1990), *New technology and the quality of work: the organizational factor*, in: Haslegrave, Wilson, Corlett and Manenica (eds.), pp. 249-258.

Fry, L.W. (1982), 'Technology-Structure Research: Three Critical Issues', *Academy of Management Journal*, 25(3), pp. 532-552.

Gahan, P. (1991), 'Forward to the Past? The Case of 'New Production Concepts'', *Journal of Industrial Relations*, 33(2), pp. 155-177.

de Galan, C., M.R. van Gils and P.J. van Strien (eds.) (1980), *Humanisering van de arbeid*, Van Gorcum, Assen.

Gallie, D. (1978), *In search of the new working class; Automation and social integration within the capitalist enterprise*, Cambridge University Press, Cambridge.

Gerlach, F. (1990), 'Facharbeitermangel - Resultat betrieblicher Einsatzpolitik ?', *WSI Mitteilungen*, 43(3), pp. 165-170.

Gorbachev, V., V. Lipping and A. Dryomov (1991), 'Perestroika And Off-line Welding', *Industrial Robot*,18(4), pp. 24-27.

Gottschalch, H. (1990), 'Produktionsarbeit in CIM-Strukturen', *Technische Rundschau*, 82(26), pp. 28-39.

Groep Sociotechniek (1986), *Het flexibele bedrijf; Integrale aanpak van flexibiliteit, beheersbaarheid, kwaliteit van de arbeid en produktie-automatisering*, Kluwer Bedrijfswetenschappen, Deventer.

Grootings, P., B. Gustavsen and L. Héthy (eds.) (1989), *New Forms of Work Organization in Europe*, Transaction Publishers, New Brunswick\London.

Groover, M.E. (1987), *Automation, Production Systems, and Computer Integrated Manufacturing*, Prentice Hall, Englewood Cliffs.

Gunn, T.G. (1987), *Manufacturing for Competitive Advantage; Becoming a World Class Manufacturer*, Ballinger, Cambridge.

Hackman, J.R. and G.R. Oldham (1980), *Work Redesign*, Addison-Wesley, Reading.

Hagedoorn, J. (1988), *Industry Structure, Innovation and Automated Manufacturing*, FAST-paper 216, Commission of the European Communities, Brussels.

Haney, L.H. (1936), *History of Economic Thought* (third edition), The Macmillan Company, New York.

Haring, H.A. and R. de Mare (1991), *Beroepspraktijk en scholingsbehoefte; Industrieel robotgebruik*, Stichting Opleidingen Metaal, Woerden.

Haslegrave, C.M., J.R. Wilson, E.N. Corlett and I. Manenica (eds.) (1990), *Work design in practice*, Taylor and Francis, London\New York\Philadelphia.

ten Have, K. (1988), 'Kwalificering en techniek in de metaalindustrie: Hoe complexe problemen simpelworden opgelost', *Tijdschrift voor Arbeidsvraagstukken*, 4(4), pp. 62-71.

ten Have, K. and B. Fruytier (1987), 'De invoering van flexibele bewerkingsmachines', *Tijdschrift voor Arbeidsvraagstukken*, 3(1), pp. 27-37.

den Hertog, J.F. (1975), *Werkstructurering*, Samsom, Alphen aan de Rijn\Brussels.

den Hertog, J.F. (1977\1979), *New Leads in Job Design: The Philips' Case*, in: Davis and Taylor (eds.),pp. 148-160.

den Hertog, F. (1990), *De lerende organisatie: lessen uit Zweden*, in: Buitendam, Dumas and Glebbeek (eds.), pp. 123-130.

den Hertog, J.F. (1991), 'Book review: David A. Buchanan and James C. Calman: High Performance Work Systems - The Digital Experience', *Organization Studies*,

12(1), pp. 141-143.

den Hertog, J.F. and F.M. van Eijnatten (eds.) (1990), *Management van Technologische Vernieuwing*, Van Gorcum, Assen\Maastricht.

Herzberg, F., B. Mausner and B. Bloch Snyderman (1959), *The Motivation to Work*, Wiley\Chapman & Hall, New York\London.

Herzog, H.-H. (1991), 'Werkstattorientierte Programmierverfahren; WOP auf dem steinigen Weg zur Norm', *Technische Rundschau*, 83(22), pp. 48-53.

van den Heuvel, N. (ed.) (1991), *New Technologies and the Future of Work*, SISWO, Amsterdam.

Hofstede, G. (1980), *Culture's Consequences; International Differences in Work-related Values*, Sage, Beverly Hills\London.

Hofstede, G. (1990), *Cultures and Organization; Software of the Mind*, McGraw-Hill, London.

Hofstede, G. and M.H. Bond (1986), 'The Confucius Connection, From Cultural Roots to Economic Growth', *Organizational Dynamics*, 16(4), pp. 4-21.

Hollier, R.H., R.J. Boaden and S.J. New (eds.) (1992), *International Operations; Crossing Borders in Manufacturing and Service*, Elsevier, Amsterdam.

Hollingum, J. (1991), 'Robots And Software At Weldex '91', *Industrial Robot*, 18(4), pp. 15-17.

van Hoof, J.J. and R. Huiskamp (1989), *New Forms of Work Organization in the Netherlands*, in: Grootings, Gustavsen and Héthy (eds.), pp. 155-174.

Hopwood, A.G. (1979), *Economic costs and benefits of new forms of work organisation*, in: ILO, pp. 114-145.

Hrebiniak, L.G. and W.F. Joyce (1985), 'Organizational Adaptation: Strategic Choice and Environmental Determinism', *Administrative Science Quarterly*, 30(3), pp. 336-349.

Huijgen, F. (1984), *Management en arbeidsbeleid*, ITS, Nijmegen.

Huijgen, F. (1989), *De kwalitatieve structuur van de werkgelegenheid in Nederland, deel III; Bevolking in loondienst en functiestructuur in 1977 en 1985*, OSA-voorstudie V 33, OSA, The Hague.

Huijgen, F. (1990), 'The training of workers and their occupational level; The qualitative structure of employment 1960-1985', *The Netherlands Journal of Social Sciences*, 26(2), pp. 91-111.

Huijgen, F. and F. Pot (1988), *Recent onderzoek naar automatisering en arbeidsorganisatie in Nederland*, in: Boekraad, van der Brink and van Wel (eds.), pp. 290-309.

Huijgen, F., B.J.P. Riesewijk and G.J.M. Conen (1983), *De kwalitatieve structuur van de werkgelegenheid in Nederland; Bevolking in loondienst en functieniveau-structuur in de periode 1960-1977*, Staatsuitgeverij, The Hague.

Huijgen, F. and J. Smits (1990), 'Verdringing op de Nederlandse arbeidsmarkt: fictie of werkelijkheid?',*Tijdschrift voor Arbeidsvraagstukken*, 6(2), pp. 19-21.

Hunt, J.W. (1971), *The Restless Organisation*, John Wiley and Sons Australasia Pty Ltd, Sydney\New York\London\Toronto.

Hyman, R. and W. Streeck (eds.) (1988), *New Technology and Industrial Relations*, Basil Blackwell, Oxford\New York.

IG Metall (1986), *Industrieroboter* (Aktionsmappe), IG Metall, s.l. [Frankfurt].

ILO (1979), *New forms of work organisation 2*, ILO, Geneva.

Industriële robots (1991), Directoraat-Generaal van de Arbeid, Voorburg.

Industrial Robot (1991), *IFR Statistics*, 19(1), pp. 4-6.

ISI - ISF (1987), *Flexible Manufacturing Systems and Cells in the Scope of New Production Systems in Germany*, FAST-paper 135, Commission of the European Communities, Brussels.

Jackson, J.H., C.P. Morgan and J.P.G. Paolillo (1986), *Organization Theory; A Macro Perspective for Management*, Prentice Hall, Englewood Cliffs.

Jaffe, A.J. and J. Froomkin (1968), *Technology and Jobs; Automation in Perspective*, Frederick A. Praeger, New York\Washington\London.

Jones, B. (1991), 'Book Reviews: Malcolm Warner, Werner Wobbe and Peter Brödner (eds.): Technology and Manufacturing Management: Strategic Choices for Flexible Production Systems and L. Bainbridge and S. Antonio Ruiz Quintanill (eds.): Developing Skills with Information Technology', *Organization Studies*, 12(4), pp. 617-619.

Kaplan, R.S. (1983), 'Measuring Manufacturing Performance; A New Challenge for Management Accounting Research', *The Accounting Review*, 68(4), pp. 686-705.

Karasek, R.A. (1979), 'Job Demands, Job Decision Latitude and Mental Strain; Implications for Job Redesign', *Administrative Science Quarterly*, 24(2), pp. 285-308.

Karlsson, C. and S. Rubenowitz (eds.) (1991), *Management and New Production Systems; The 3rd International Production Management on Management and New Production Systems*, EIASM\University of Göteborg, Brussels\Göteborg.

Karlsson, J.M. (1991), *A Decade of Robotics; Analysis of the Diffusion of Industrial Robots in the 1980s by Countries, Application Areas, Industrial Branches and Types of Robot*, Mekanförbundets Förlag, Stockholm.

Kast, F.E. and J.E. Rosenzweig (1981), *Organization and Management; A Systems and Contingency Approach* (third edition), McGraw-Hill, Tokyo.

Keefe, J.H. (1991), 'Numerically Controlled Machine Tools and Worker Skills', *Industrial and Labor Relations Review*, 44(3), pp. 503-519.

Kelley, M.R. (1990), 'New Process Technology, Job Design, and Work Organization: A Contingency View', *American Sociological Review*, 55(1), pp. 191-208.

Kern, H. and M. Schumann (1970), *Industriearbeit and Arbeiterbewußtsein; Eine empirische Untersuchung über den Einfluß der aktuellen technischen Entwicklung auf die industrielle Arbeit und das Arbeitersbewußtsein*, Europäische Verlagsanstalt, Frankfurt am Main.

Kern, H. and M. Schumann (1984), *Das Ende der Arbeitsteilung ?'; Rationalisierung der industriellen Produktion*, Beck, Munich.

Kern, H. and M. Schumann (1987), 'Limits of the Division of Labour. New Production and Employment Concepts in the West German Industry', *Economic and Industrial Democracy*, 8(2), pp. 151-170.

Kerr, C., J.T. Dunlop, F.H. Harbison and C.A. Myers (1960 (1962), *Industrialism and Industrial Man; The Problems of Labor and Management in Economic Growth*, Heinemann, London.

Khalil, T.M. and B.A. Bayraktar (eds.) (1992), *Management of Technology III; The Key to Global Competitiveness*, Industrial Engineering and Management Press,

Norcross.

Kidd, P.T. (1991), 'Human and Computer Integrated Manufacturing: A Manufacturing Strategy Based on Organization, People, and Technology', *International Journal of Human Factors in Manufacturing*, 1(1), pp. 17-32.

Kieser, A. (1988), *Adapting Organizations to Micro-Electronics; Experiences in the Federal Republic of Germany*, in: Urabe, Child and Kagono (eds.), pp. 319-336.

Kloosterman, R.C. and T. Elfring (1991), *Werken in Nederland*, Academic Service, Schoonhoven.

KNOV (1991), *Vernieuwen doe je niet alleen; Naar een meer op het mkb gericht technologiebeleid*, KNOV, Rijswijk.

Koch, J. (1988), 'Brave Old World; De goede bedoelingen van Frederick Winslow Taylor', *Intermediair*,24(36), pp. 63-65.

Kommers, H. and A. Maas (1987), *De gevalstudie; Een oriënterende benadering*, IVA, Tilburg.

Koopman-Iwema, A.M. (ed.) (1986), *Reorganiseren is automatiseren; Richtlijnen voor het personeelsmanagement*, Kluwer\NVP, Deventer.

de Korte, D.J. (1991), 'NC-Programmering; Ondersteuning bij de keuze van een NC-programmeersysteem', *MB*, 57(2), pp. 50-61.

Krabbendam, J.J. (1988), *Nieuwe technologieën en organisatorische maatregelen*, unpublished PhD-dissertation University of Twente, s.n. [Febodruk], s.l. [Enschedé].

Krabbendam, J.J. and J.A.M. Willenborg (1986), 'Organisatorische voorwaarden voor flexibele fabricagesystemen in de metaalindustrie', *M&O*, 40(5), pp. 423-436.

Kuipers, H. and P. van Amelsvoort (1990), *Slagvaardig organiseren; Inleiding in de sociotechniek als integrale ontwerpleer*, Kluwer Bedrijfswetenschappen, Deventer.

Kuipers, J.H. (1972), *Verantwoordelijkheidsverruiming in de direkte werksituatie*, SWI-VU\COP-SER, Amsterdam\The Hague.

Kuylaars, A.M. (1951), *Het verband tussen werk en leven van de industriële loonarbeider, als object van een sociale ondernemingsproblematiek*, Stenfert Kroese, Leiden.

Lambrinos, J. and W.G. Johnson (1984), 'Robots to reduce the high cost of illness and injury', *Harvard Business Review*, 62(3), pp. 24,26,28.

Lammers, C.J. (1983), *Organisaties vergelijkenderwijs; Ontwikkeling en relevantie van het sociologisch denken over organisaties*, Het Spectrum, Utrecht.

Lammers, C.J. (1986), 'De excellente onderneming als organisatiemodel', *Harvard Holland Review*, 8,pp. 18-27.

Lammers, C.J. (1990), 'Sociology of Organizations Around the Globe; Similarities and Differences Between American, British, French, German and Dutch Brands', *Organization Studies*, 11(2), pp. 179-205.

Landau, K. and U. Brandenburg (1986), 'Zur Analyse von Anforderungsverschiebungen durch Einsatz von Industrierobotern im Automobilbau', *Zeitschrift für Arbeitswissenschaft*, 42(4), pp. 201-205.

Lane, C. (1989), *Management and Labour in Europe; The Industrial Enterprise in Germany, Britain and France*, Edward Elgar, Aldershot.

Langton, J. (1984), 'The Ecological Theory of Bureaucracy: The Case of Josiah Wedgwood and the British Pottery Industry', *Administrative Science Quarterly*,

29(3), pp. 330-354.

Lauenstein, T., D. Seitz and V. Volkholz (1989), *Industrieroboter im Routinebetrieb; Arbeits- und sozialwissenschaftliche Analysen in heterogenen Einsatzfeldern*, Campus Verlag, Frankfurt\New York.

Laurentius, G., H. Timmerman and A.A.M. Vermeulen (1982), *Flexibele automatisering in Nederland; Ervaringen en opinies*, Delftse Universitaire Pers, Delft.

Lee, D. (1982), *Beyond deskilling: skill, craft and class*, in: Wood (ed.), pp. 146-162.

Leermakers, P.G.J.L. (1991), *Industriële robots en de werkverdeling*, unpublished Master's thesis TilburgUniversity, s.n., Tilburg.

Lenin, W.I. (1914\1961), 'Das Taylorsystem - die Versklavung des Menschen durch die Maschine', *WI Lenin Werke*, Dietz Verlag, Berlin.

Lenin, W.I. (1918\1960), 'Die nächsten Aufgaben der Sowjetmacht', *WI Lenin Werke*, Dietz Verlag, Berlin.

Locke, E.A. (1982), 'The Ideas of Frederic W. Taylor: An Evaluation', *Academy of Management Review*, 7(1), pp. 14-24.

Lok, N.R. and J. Slomp (1990), 'De ontwikkeling van een produktiebesturingssysteem voor een FFS; Een case-beschrijving', *MB*, 56(10), pp. 356-363.

Loontechnische Dienst (1991), *Functiewaardering in Nederland 1989*, Ministry of Social Affairs and Employment, The Hague.

MacConaill, P.A., P. Drews and K.-H. Robrock (eds.) (1991), *Mechatronics & Robotics I*, IOS Press, Amsterdam\Oxford\Washington\Tokyo.

Manders, A.J.C. (1990), *Sturing van produktie-technologie; Een sociaal-economische case-studie over fabricage-technologie*, unpublished PhD-dissertation University of Utrecht, s.n. [Kerckebosch], s.l. [Zeist].

Mangold, V.L. (1991), 'Safety and Robotic Welding: Mutual Compatibility', *Welding Journal*, 70(9), pp. 71-74.

Manske, F. (1987), *Alternative Strategies of Production Planning and Control*, in: Brödner (ed.), pp. 131-140.

March, J.G. and H.A. Simon, *Organizations*, Wiley, New York.

Marshall, A. (1890\1936), *Principles of Economics; An introductory volume* (reprint of the eight edition of1920), Macmillan, London.

Marx, K. (1867\1932), *Das Kapital; Das Produktionsprozess des Kapitals*, Gustav Kiepenheuer Verlag, Berlin.

Marx, K. and F. Engels (1848\1945), *Het communistisch manifest*, Pegasus, Amsterdam.

Mason, G., S.J. Prais and B. van Ark (1992), 'Vocational Training and Productivity in the Netherlands and Britain', *National Institute Economic Review*, 140, pp. 45-63.

Maurice, M., A. Sorge and M. Warner (1980), 'Societal Differences in Organizing Manufacturing Units, A Comparison of France, West Germany and Great Britain', *Organization Studies*, 1(1), pp. 59-86.

Maynard, H.B., G.J. Stegemerten and J.L. Schwab (1948), *Methods-Time Management*, McGraw-Hill: New York\Toronto\London.

McCann, J.E. and J. Selsky (1984), 'Hyperturbulence and the Emergence of Type 5 Environments', *Academy of Management Review*, 9(3), pp. 460-470.

Mickler, O., E. Dittrich and U. Neumann (1976), *Technik, Arbeitsorganisation und*

Arbeit; Eine empirische Untersuchung in der automatisierten Produktion, Aspekte Verlag, Frankfurt am Main.

Mickler, O., W. Pelull, W. Wobbe-Ohlenburg, P. Kalmbach, R. Kasiske and F. Manske (1981), *Industrieroboter; Bedingungen und soziale Folgen des Einsatzes neuer Technologien in der Automobilproduktion*, Campus Verlag, Frankfurt\New York.

Miller, J. (1988), *Jobs and Work*, in: Smelser (ed.), pp. 327-359.

Morgan, G. (1986), *Images of Organization*, Sage: Beverly Hills\Newbury Park\London\New Delhi.

Mossink, J. and M. Peeters (1992), *Robots, working conditions and job content; Opportunities and hindrances for the improvement of the quality of working life*, Preprints Fifth IFAC\ICIP\IFORS\IEA Symposium on Man-Machine Systems, s.n., The Hague, section 5.1.3: 1-8.

Mueller, F. (1992), 'Flexible Working Practices in Engine Plants: Evidence from the European Automobile Industry', *Industrial Relations Journal*, 23(3), pp. 191-204.

Mueller, F. and J. Purcell (1992), 'The Europeanization of manufacturing and the decentralization of bargaining: multinational management strategies in the European automobile industry', *International Journal of Human Resource Management*, 3(1), pp. 15-34.

Murphy, S. (1989), *The ESPRIT project*, in: Rosenbrock (ed.), pp. 145-168.

Nagamachi, M. (1986), *Ergonomics and the Introduction of Industrial Robots*, in: Noro (ed.), pp. 339-342.

Nielsen, K. (1992), 'CNC technology, job design and training: the scope for personnel strategies', *New Technology, Work and Employment*, 7(2), pp. 116-124.

de Nijs, W. (1987), *Alain Touraine; Het sociologisch geweten van de geprogrammeerde samenleving*, Tilburg University Press, Tilburg.

Noble, D. (1978), 'Social Choice in Machine Design: The Case of Automatically Controlled Machine Tools, and a Challenge for Labor', *Politics and Society*, 8(3-4), pp. 313-347.

Noro, K. (ed.) (1986), *Occupational Health and Safety in Automation*, Taylor and Francis, London\NewYork\Philadelphia.

Nullmeier, E. and K.H. Rödiger (1986), *Arbeitsorientierte Anforderungen an die Gestaltung von PPS-Systemen. Rechnerintegrierte Produktion - zur Entwicklung von Technik und Arbeit in der Metallindustrie*, Campus Verlag, Frankfurt\New York.

Okko, P. (1992), *Modern Manufacturing and Small Firms*, in: Khalil and Bayraktar (eds.), pp. 888-897.

Okubayashi, K. (1986), 'The impact of industrial robots on working life in Japan', *Journal of General Management*, 11(4), pp. 22-34.

O'Reilly, J. (1992), 'Banking on flexibility; a comparison of the use of flexible employment strategies in the retail banking sector in Britain and France', *International Journal of Human Resource Management*, 3(1), pp. 35-58.

OSA (1990), *Arbeidsmarktperspectieven; OSA Rapport 1990*, OSA, The Hague.

OSA (1992), *Arbeidsmarktperspectieven; OSA Rapport 1992*, OSA, The Hague.

Peeters, M.H.H. and F.D. Pot (with cooperation of N.J. Delleman) (1991), *Nieuwe produktiemethoden voor de kledingindustrie; Een verkennend onderzoek naar de mogelijkheden van integratie van het aspect kwaliteit van de arbeid in een (integraal) herontwerp*, NIPG publication 91.015, NIPG-TNO, Leiden.

Perrow, C. (1967a), 'A Framework for the Comparative Analysis of Organizations', *American Sociological Review*, 32(2), pp. 194-208.

Perrow, C. (1967b), 'Book Review: Industrial Organization: Theory and Practice by Joan Woodward', *American Sociological Review*, 32(3), pp. 313-315.

Perrow, C. (1983), 'The Organizational Context of Human Factors Engineering', *Administrative Science Quarterly*, 28(4), pp. 521-541.

Perrow, C. (1986), *Complex Organizations; A Critical Essay* (third edition), Random House, New York.

Peters, R. (1986), *De inzet van CNC-bewerkingsmachines in de produktie en de kwaliteit van de arbeid; Een verkennend onderzoek in zes bedrijven naar situatie, ervaringen en mogelijke invloedsfaktoren*, unpublished Master's thesis Eindhoven University of Technology, Eindhoven.

Peters, R. and J. Kommers (1991), *Produktiebesturing, automatisering en kwaliteit van werk en organisatie*, IVA, Tilburg.

Piore, M. and C. Sabel (1984), *The Second Industrial Divide; Possibilities for Prosperity*, Basic Books, New York.

Popitz, H., H.P. Bahrdt, E.A. Jüres and H. Kesting (1957), *Technik und Industriearbeit; Soziologische Untersuchungen in der Hüttenindustrie*, J.C.B. Mohr, Tübingen.

Porter, M.E. (1985), *Competitive Advantage; Creating and Sustaining Superior Performance*, The Free Press, New York.

POST-groep (1991), *Onderweg naar nieuwe fabrieken en kantoren*, Kluwer Bedrijfswetenschappen, Deventer.

Pot, F.D., M.H.H. Peeters, P. van Amelsvoort and J. Middendorp (1991), *Functieverbetering en integraal ontwerpen; Richtlijnen voor het verbeteren van het aspect welzijn bij de arbeid in ontwerpprojecten*, NIPG publication 91.010, NIPG-TNO, Leiden.

Poutsma, E. (1991), *Technique, organization and the future of work in manufacturing*, in: van den Heuvel (ed.), pp. 39-54.

Poutsma, E., P.M. van der Staal and A.B. Zwaard (1986), *Automatisering: effecten voor het midden- en kleinbedrijf*, EIM, Zoetermeer.

Poutsma, E. and A. Zwaard (1987), 'Automatisering en arbeidsinhoud; Effecten van CNC-apparatuur in kleine, industriële bedrijven', *Tijdschrift voor Arbeidsvraagstukken*, 3(1), pp. 16-26.

Poutsma, E. and A. Zwaard (1989), 'Programming CNC-equipment-The effects of automation in small industrial enterprises', *International Small Business Journal*, 7(2), pp. 35-43.

Projectgroep WEBA (1989), *Functieverbetering en kwaliteit van de arbeid; Welzijn bij de arbeid (WEBA) gelet op de stand van de arbeids- en bedrijfskunde*, Ministry of Social Affairs and Employment, Voorburg.

Ramondt, J.J. (1968), *Verantwoordelijkheid in het werk; Verslag van een experiment in zes industriële bedrijven*, Samsom, Alphen aan de Rijn\Brussels.

Ramondt, J. (1974), *Bedrijfsdemocratisering zonder arbeiders; Een evaluatie van ervaringen met werkoverleg en werkstructurering*, Samsom, Alphen aan de Rijn\

Brussels.

Rauwenhoff, F.C. (1986), 'De fabriek van de toekomst; Een toekomstgerichte industriële strategie', *Economisch Statistisch Berichten*, 71(3538), pp. 32-35.

Reijers, L.N. (1992), 'Automatisering in de produktie; Evolutie of revolutie', *MB*, 58(2), pp. 41-48.

Reijers, L.N. and H.J.L.M. de Haas (1986), *Flexibele Produktie-automatisering; Industriële robots; (deel III)*, De Vey Mestdagh, Middelburg.

Ricardo, D., 1817 (1911), *The Principles of Political Economy & Taxation*, J.M. Dent & Sons, London\Toronto.

Robbins, S.P. (1984), *Management; Concepts and Practices*, Prentice Hall, Englewood Cliffs.

Robots in Japan; Flexibele automatisering in de produktie (1982), Staatsuitgeverij, Den Haag.

Rogers, E.M. (1962), *Diffusion of Innovations*, Free Press\Collier-MacMillan, New York\London.

Roobeek, A. (1987), *Robots en flexibele automatisering in de auto-industrie*, in: Steensma and Vrooland (eds.), pp. 86-98.

Rosenbrock, H.H. (ed.) (1989), *Designing Human-Centred Technology; Across Disciplinary Project Computer-aided Manufacturing*, Springer Verlag, London.

Routh, G. (1989), *The Origin of Economic Ideas* (second edition), MacMillan, Basingstoke.

Rubenstein, A.H. and C.J. Haberstroh (eds.) (1966), *Some Theories of Organization* (revised edition), Irwin-Dorsey, Homewood.

Rumberger, R.W. (1981), 'The Changing Skill Requirements of Jobs in the U.S. Economy', *Industrial and Labor Relations Review*, 34(4), pp. 578-591.

Salverda, W. (1990), 'Verdringing op de Nederlandse arbeidsmarkt', *Tijdschrift voor Arbeidsvraagstukken*, 6(2), pp. 4-18.

Scherff, B. (1989), 'Ergonomischer Einsatz von Sprache als Teil der Mensch-Maschine-Schnittstelle zur Programmierung von CNC-Robotern für das Schweißen', *Zeitschrift für Arbeitswissenschaft*, 15(4), pp. 224-233.

Schonberger, R.J. (1982), *Japanese Manufacturing Techniques - Nine Hidden Lessons in Simplicity*, The Free Press, New York.

Schumann, M., V. Baethge-Kinsky, U. Neumann and R. Springer (1990), 'Breite Diffusion der Neuen Produktionskonzepte - zögerlicher Wandel der Arbeitsstrukturen; Zwischenergebnisse aus dem "Trendrapport - Rationalisierung der Industrie"', *Soziale Welt*, 41(1), pp. 46-69.

Schünemann, T.M. and T. Bruns (1986), 'Modellierung und Prognose der Diffusion von Industrierobotern in der Bundesrepublik Deutschland', *Zeitschrift für Betriebswirtschaft*, 56(10), pp. 953-988.

Shaiken, H., S. Herzenberg and S. Kuhn (1986), 'The Work Process Under More Flexible Production',*Industrial Relations*, 25(2), pp. 167-183.

Sismondi, J.-C.L. Sismonde de (1819\1971), *Nouveaux principes d'économie politique ou De la richesse dans ses rapports avec la population*, Calmann-Lévy, Paris.

de Sitter, L.U. (1974a), 'Sociotechniek', *M&O*, 28(2), pp. 65-83.

de Sitter. L.U. (1974b), 'Sociotechniek 2; Aantekeningen bij een sociotechnisch model van een produktiesysteem', *M&O*, 28(3), pp. 163-176.

de Sitter, L.U. (1981), *Op weg naar nieuwe fabrieken en kantoren*, Kluwer, Deventer.

de Sitter, L.U. (1989), *Modern Sociotechnology*, KOERS Consultants, Den Bosch.

de Sitter, U. and F. den Hertog (1990), *Simple Organisations, Complex Jobs, The Dutch Sociotechnical Approach*, Research Memorandum 90-012, MERIT, Maastricht.

Skinner, W. (1966), 'Production under pressure', *Harvard Business Review*, 44(6), pp. 139-146.

Slomp, J. and G.J.C. Gaalman (1992), *A Production Control System for a Flexible Manufacturing System - Description and Evaluation*, in: Eyada and Ahmad (eds.), pp. 654-665.

Slomp, J. and J.N.D. Gupta (1992), 'An Interactive Tool for Scheduling Jobs in a Flexible Manufacturing Environment', *International Journal of Computer Integrated Manufacturing Systems*, forthcoming.

Slomp, J., A.T. Schipper and H. Kroon (1990), 'Plaatsingsmogelijkheden van FFS in de Nederlandse metaalindustrie; Deel 2: De markt voor flexibele fabricagesystemen', *MB*, 56(1), pp. 30-35.

Slomp, J. and J.C. Spijkerman (1988), 'Het succes van een flexibel fabricagesysteem', *MB*, 54(10), pp. 380-384.

Slomp, J. and J.D. Verbeek (1989), 'De gevolgen van een flexibel fabricagesysteem voor de produktiebesturing; Een case-beschrijving', *MB*, 55(10), pp. 348-353.

Slomp, J., M.F. de Weerd, A.T. Schipper and H. Kroon (1989), 'Plaatsingsmogelijkheden van FFS in de Nederlandse metaalindustrie; Deel 1: Een individuele geschiktheidsbeoordeling', *MB*, 55(12), pp. 431-436.

Slomp, J. and W.H.M. Zijm (1992), 'A Manufacturing Planning and Control System for a Flexible Manufacturing System', *Robotics and CIM*, forthcoming.

Smelser, N.J. (ed.) (1988), *Handbook of Sociology*, Sage, Newbury Park.

Smith, A. (1776\1976), *An Inquiry into the Nature and Causes of the Wealth of Nations* (Cannan edition), University of Chicago Press, Chicago.

Sombart, W. (1902), *Der Moderne Kapitalismus; Zweiter Band. Die Theorie der Kapitalistischen Entwicklung*, Duncker & Humblot, Leipzig.

Sørensen, K.N. (1985), 'Technology and Industrial Democracy; An inquiry into some theoretical issues and their social basis', *Organization Studies*, 6(2), pp. 139-160.

Sorge, A. (1985), *Informationstechnik und Arbeit im sozialen Prozeß*, Campus, Frankfurt\New York.

Sorge, A. (1989), 'An Essay on Technical Change: Its Dimensions and Social and Strategic Context', *Organization Studies*, 10(1), pp. 23-44.

Sorge, A. (1991), 'Strategic Fit and the Societal Effect: Interpreting Cross-National Comparisons of Technology, Organization and Human Resources', *Organization Studies*, 12(2), pp. 161-190.

Sorge, A. and W. Streeck (1988), *Industrial Relations and Technical Change: The Case for an Extended Perspective*, in: Hyman and Streeck (eds.), pp. 19-47.

Sorge, A. and M. Warner (1986), *Comparative Factory Organisation; An Anglo-German Comparison of Manufacturing*, Gower, Aldershot.

Spenner, K.I. (1983), 'Deciphering Prometheus: Temporal Change in the Skill Level of Work', *American Sociological Review*, 48(6), pp. 824-837.

Stecke, K.E. and R. Suri (eds.) (1989), *Proceedings of the Third ORSA/TIMS*

Conference on Flexible Manufacturing Systems: Operations Research Models and Applications, Elsevier, Amsterdam.

Steensma, H.O. en V.C. Vrooland (eds.) (1987), *Automatiseren gaat niet vanzelf; Een overzicht van nieuwe theorie, praktische invoeringservaringen en toekomstig beleid*, Stichting CCOZ, Amsterdam.

Steijn, A.J. and de Witte, M.C. (1985), 'Technologie en arbeid; De tendens naar steeds hogere kwaliteits-en kwalificatie-eisen', *Economisch Statistische Berichten*, 70(3530), pp. 1106-1110.

Stokman, C.T.M., A.G.A. de Groot and R.W.C. Leurs (1989), *Flexibele produktie-automatisering in midden- en kleinbedrijf*, EIM, Zoetermeer.

Sydow, J. (1985), *Der soziotechnische Ansatz der Arbeits- und Organisations-gestaltung; Darstellung, Kritik, Weiterentwicklung*, Campus Verlag, Frankfurt-\New York.

Tani, A. (1992), *The Diffusion of Robots*, in: Ayres, Haywood and Tchijov (eds.), pp. 155-170.

Taylor, F.W. (1903\1911), *Shop Management*, Harper and Brothers Publishers, New York\London.

Taylor, F.W. (1911\1914), *The Principles of Scientific Management*, Harper and Brothers Publishers, New York\London.

Tchijov, I. (1992), *The Diffusion of Flexible Manufacturing Systems*, in: Ayres, Haywood and Tchijov (eds.), pp. 197-248.

Teulings, C. (1990), 'De verschuiving in de relatie tussen beroep en opleiding: substitutie of verdringing?',*Tijdschrift voor Politieke Ekonomie*, 13(4), pp. 82-93.

Thompson, P. (1989), *The Nature of Work; An Introduction to Debates on the Labour Process* (second edition), MacMillan, Basingstoke.

Tidd, J. and A. Francis (1992), *Lean Production and Continuous Product Development: The Productivity Dilemma Revisited*, in: Khalil and Bayraktar (eds.), pp. 807-816.

Touraine, A. (1955), *L'évolution du travail ouvrier aux usines Renault*, Centre National de la Recherche Scientifique, s.l. [Paris].

Trist, E.L. and K.W. Bamforth (1951), 'Some social and psychological consequences of the longwall method of coal-getting', *Human Relations*, 4(1), pp. 3-38.

Urabe, K., J. Child and T. Kagono (eds.) (1988), *Innovation and Management; International Comparisons*, de Gruyter, Berlin.

Urban, G. and K. Hölldampf (1988), *Menschengerechte Arbeitsgestaltung beim Schweißen mit Robotern; Beispielhafte Lösungsmöglichkeiten für verschiedene Einsatztypen*, Springer-Verlag, Berlin\Heidelberg\New York\London\Paris\Tokyo.

van der Vegt, I. (1992), *De arbeidsverdeling rond booglasrobots*, unpublished Master's thesis Tilburg University, s.n., Tilburg.

Verbraeken, A.J.G. (1982), *Flexibele fabricage-systemen*, in: Robots in Japan, pp. 38-52.

Vermeulen, A.A.M. (1985), 'Flexibele produktie-automatisering: het gebruik van industriële robots', *Informatie*, 27(special issue), pp. 407-414.

Vermeulen, I. and R. van Veen (1991), *Robots in Europe; the robotpopulation at 31 december 1990; statistic report* (Edition: Benelux), ANERTEK, Bilthoven.

Vickery, G. and D. Campbell (1989), 'Advanced Manufacturing Technology and the

Organisation of Work', *STI Review*, 6, pp. 105-154.

Vissers, A., G. Evers, M. van Tits, H. Vermeulen and R. Euwals (1991), *De ontwikkelingen op de arbeidsmarkt in de zorgsector tot het jaar 2000*, IVA, Tilburg.

Vollmann, T.E., W.K. Berry and D.C. Whybark (1984), *Manufacturing Planning and Control Systems*, Irwin, Homewood.

Vrakking, W.J. (1988), *Zaken doen met FPA; Adviseren en leveren bij flexibele produktie-automatisering*, Kluwer\NEHEM, s.l. [Deventer].

van der Waerden, T. (1911), *Geschoolheid en techniek; Onderzoek naar den invloed van arbeidssplitsing en machinerie op de mate van vereischte oefening en bekwaamheid des arbeiders*, F. van Rossum, Amsterdam.

van der Waerden, T. (1916), *Het Taylorstelsel; Met een inleiding over stukloon en moderne loonsystemen*, Ontwikkeling, Amsterdam.

Walker, C.R. (1950), 'The Problem of the Repetitive Job', *Harvard Business Review*, 28(3), pp. 54-58.

Walker, C.R. and R.H. Guest (1952), *The Man on the Assembly Line*, Harvard University Press: Cambridge.

Weber, M. (1966), *Bureaucracy*, in: Rubenstein and Haberstroh (eds.), pp. 70-81.

Weinstein, D. (1979), *Bureaucratic Opposition; Challenging Abuses at the Workplace*, Pergamon Press, New York\Oxford\Toronto\Sydney\Frankfurt\Paris.

Wexley, K.N. and G.A. Yukl (1984), *Organizational Behavior and Personnel Psychology* (revised edition), Irwin, Homewood.

Wielers, R. and A. Glebbeek (1990), 'Worden we écht te slim voor ons werk ? Drie interpretaties van de onderzoeksresultaten van Huijgen', *Mens en Maatschappij*, 65(3), pp. 271-288.

Willenborg, J.A.M. (1987), *Flexibele automatisering en organisatie; Veranderingen in de organisatie bij invoering van flexibele fabricage-systemen*, unpublished PhD-dissertation University of Twente, s.n. [TISO], s.l. [Enschedé].

Willenborg, J.A.M. (1990), *Flexibiliteit van automatisering en organisatie; een paradox en dilemma*, in: den Hertog and van Eijnatten (eds.), pp. 109-128.

Willenborg, J.A.M. and J.J. Krabbendam (1987), 'Industrial automation requires organizational adaptations', *International Journal of Production Research*, 25(11), pp. 1683-1691.

Wilson, F.M. and D.A. Buchanan (1988), 'The Effect of New Technology in the Engineering Industry: Cases of Control and Constraint', *Work, Employment & Society*, 2(3), pp. 366-380.

Winch, G. (ed.) (1983), *Information Technology in Manufacturing Processes; Case Studies in Technological Change*, Rossendale, London.

Wobbe, W. (1989), *Employments effects of robots and flexible manufacturing*, FAST-paper 218, Commission of the European Communities, Brussels.

Wobbe, W. (1991), *Anthropocentric Production Systems; A Strategic Issue for Europe*, FAST-paper 245, Commission of the European Communities, Brussels.

Womack, J.P., D.T. Jones and D. Roos (1990), *The Machine that Changed the World*, Rawson & Collier, New York.

Wood, S. (ed.) (1982), *The Degradation of Work ? Skill deskilling and the labour process*, Hutchinson, London\Melbourne\Sydney\Auckland\Johannesburg.

Woodward, J. (1965), *Industrial Organization; Theory and Practice*, Oxford University Press, London.

Yin, R.K. (1989), *Case Study Research; Design and Methods* (second edition), Sage Publications, Newbury Park\London\New Delhi.

Yokomizo, Y. (1986), *Occupational Health and Safety Problems Caused by Increases in the Robot Population*, in: Noro (ed.), pp. 169-174.

van der Zwaan, A.H. (1975), 'The Sociotechnical Systems Approach; A Critical Evaluation', *International Journal of Production Research*, 13(2), pp. 149-163.

van der Zwaan, A.H., A.A.M. Vermeulen, P. Smets and E. Elzinga (1974), 'Sociotechniek 4; Onderzoeksmethoden inzake de produktiebesturing', *M&O*, 28(6), pp. 342-369.

van Zwol, W.I. and A. Bots (1992), *Kosten van de primaire vakopleidingen in de metaalsector; Een studie naar de kostenstructuur van de primaire vakopleidingen in het kader van het leerlingwezen bij de bedrijven in de metaalsector*, EIT, Tilburg.

Author index